Super Controller Guide

Super *Controller Series Book 1, 2, and 3*
For Dealership Management

BY
SANDI JEROME
Author of the Profit Accounting Guide

SmilingEagle Press Book

Published by SmilingEagle Press
For information:
SmilingEagle Press
www.smilingeagle.com

ISBN: 978-1-969767-11-1

Printed in the United States of America
Illustrations by Tulaasi Jerome

CONTENTS

About the Author

Sandi Jerome started in the car business when she was fourteen keeping the parts inventory on 3×5 cards. She was one of the first female F&I managers and eventually became a CFO, fixed operations manager, and assistant general manager for the largest Chevrolet and Dodge dealer in a 5-state area. She helped start and was the first editor of ***Digital Dealer Magazine***. For many years, she wrote a monthly column for the magazine and often spoke at the Digital Dealer Conference and the NADA convention. Her workshops, “40 Valuable Things to Do with Your Computer System” and “How to Create a Super Controller” were ranked in the top 10 at NADA. She is CPA with a degree in accounting and minor in computer science. She was the first certified ADP/CDK consultant and helped ADP write their certification program. She has worked on over 20 different DMS systems, helping “Americanize” two foreign systems for the US market and teaching other DMS companies and 3rd parties how to extract data from ADP/CDK, R+R and other DMS systems. Her consulting firm, Sandi Jerome Computer Consulting helped over 1,100 dealers pick between 20 different DMS systems and eventually she created her own DMS, DealerStar DMS. Over the years, DealerStar has been installed in over 15 states four countries and their largest single point client has 7 franchises and almost 200 users on the DealerStar DMS system. Their enterprise design enables a global company with almost 60 dealerships in the US to operate on DealerStar since 2019. After obtaining GM, KIA, VW, Audi, Ford, Honda and Acura certification, DealerStar merged with DealerTeam, Inc. a major CRM provider developed on the Salesforce platform. Sandi retired from the industry in 2022 and is now a full-time writer.

Dedication

So many people helped me on my journey, including some fantastic dealers who taught me a lot about this business. But the two that I appreciate most are Robert Jean-Louis and Jackie Hartgrove. Without them, I would not have been able to turn over the management of DealerStar DMS and enable my retirement from the industry – and allow me the time to write full-time. But the special dedication is to my husband, Keith Jerome. He gave me confidence, strength, and, most importantly, the time to become a Super Controller. I rewrote this book on the eve of our 53rd wedding anniversary and all I can say now --- it has been a great ride. *More adventures Sancho Panza!*

Note from Author

Yes, this is the same guide that I'd sell for $195-$395 while I was a consultant. That is because it came with unlimited support from me. My typical day began with about forty emails from controllers, office manager, dealers and general managers with questions, "How do I…?"

It was also a pain to make the CD, print, bind and mail out. I solved that when I started to retire by putting it up on Amazon, but I couldn't price it like a normal book. What about someone who a few years ago paid $395 or $995 for all three? The price was $195 for all three for a few years, so a little closer.

Today, I got a call from one of my Adoma friends, asking me to speak in Las Vegas. I love this group of office managers and controllers, but my current responsibilities prevent me from going. I want to "be there" and thought of this guide. What if I trim this down a little, print all black and white, and republish it at a lower price? They could give away a few free copies out and then if I price it very low, even the cheapest dealer could buy it for their office manager or controller – OR…maybe even for the General Manager so that person could **get an inkling of what we do all day!**

I am the Queen of Shameless Self-Promotion, and this guide had references to the DMS that I designed long ago – DealerStar, that is now part of a bigger company, DealerTeam. I'm editing a lot of that out but taking this moment to mention that there is an afterlife. Yup, we can retire from this business. While I was a controller, CFO, then owner of DealerStar, I knew what I was going to do when I no longer dreaded the month-end close. I wanted to write full-time and I'm living that dream.

If you visit my website at SandraJerome.com – you'll see that I've been busy these past few years. My specialty is adapting best-selling books into screenplays and some of those are now in production. I'm also learning AI Animation and making some web series with my own books.

If you want to see more, just Google me (ohh..that tickles!) and search on Sandi Jerome YouTube – and you can see what I'm working on along with shameless self-promotion of my two Young Adult series, ***Wilma Wallaby*** and ***Kira and Henry***. Either might be perfect for your General Manager or dealer to read – since they are written at about 5th grade level. Okay…couldn't resist…Enjoy, Survive, Thrive!

Your friend - Sandi

Super Controller Guide - Introduction

Are you ready to become a "Super Controller or Super CFO?" What is the difference between an office manager and a Super Controller? A Super Controller cares about the total dealership – not just the accounting office. How do learn about the parts department, service department and sales without spending years working in those departments? The answer is this guide!

This guide is actually three books; Super Controller, Super Controller 2 – Employees and Expenses and Super Controller 3 – Time and Tasks. My name is Sandi Jerome I like to think that I was a good – even "Super" Controller. I did all the things that showed that I was "in control" of my accounting office. I had super clean schedules and always closed the books in 3 working days. I was an expert on the DMS system; I ended up being an expert on over a dozen DMS systems which made me a perfect person to design a new DMS system. I enjoyed writing reports to dazzle my dealer and custom the system when possible. In particular, I liked to provide detailed reports that that pointed out every mis-step by my fellow managers; a daily heat sheet, past due accounts, over-aged inventory. I guess I liked to set my fellow managers on fire and then throw gasoline on fire!

As a Super Controller, I was prepared to work in any department; service, parts, and sales. I had been one of the first female F&I mangers. Then as a controller, my special talent was that I trained GSMs to be GMs and eventually became an assistant General Manager. In my organization, I was always ready to do more work. One day, I got the "call!" I was sure that they were going to ask me to be the GM of a new dealership they were buying. Instead, when I reported to the corporate offices, they had more reconciliation work for me to do. After completing that, I asked about the new dealership and they set me up with a meeting with the "big guy" in charge. I sat across from him later that week, resume in hand and he merely leaned back and said, "I've put "bean counters" in that position before and it never works out. Bean Counter? How could I be thought of as a bean counter? Did he look at my resume? I started taking parts inventory when I was only fourteen. I was an amazing F&I manager – I won awards for high penetration and averages per deal. I knew the dealership fixed operations backwards and forward. I had trained so many GMs! How can they think of me as a bean counter?

After I sulked and returned to my dealership and desk, I analyzed the meeting in my head and was focused on the bean counter label. What had I done to deserve that? Then it hit me. I thought back to all the manager's meetings and things that I discussed. I talked about deals not in the office, paperwork missing, past due accounts; negative, negative, negative. Although I created amazing reports – did they really help managers make more profit? I was a miser when it came to expenses, but how much could I really cut? What had I done lately to increase profit? I remembered painfully one meeting where I had complained about too many staples in car deals. Really!

Later in my career I had a second chance to "reinvent" myself. In my new job, the dealer was frustrated over the bottom line in the service department. He said that if I could make it profitable, then I'd get a big bonus. Well, that is easy – just change the percentage of expense distribution! But that wasn't an option. Instead, I had to increase service gross profit. I had taken an extensive service management training course and installed dispatching software, but I had to ask myself, "how does a service department make gross? Why do parts shortages happen?" If you are coming into to this industry from what we call "the real world" and have been told over and over that "we do things differently in the car business," then this might help you understand better.

Is this guide outdated? I hope so. I hope things have changed – I certainly have!

Good luck!

Sandi

Super Controller Guide – Book One

This first book is broken down into five sections;

- The Sales Department
- The Service Department
- The Parts Department
- The Accounting Department
- The Computer System

We start in the Sales Department and work our way through service and parts and end in the accounting Department because if you streamline your daily and monthly accounting functions – you'll have the time you need to be a Super Controller. If you are new to the industry I've included the basic explanations of how things work and guidelines to improve how the dealership manages these departments. The manual can also be a reference guide and training manual for those entering the automotive industry along with the more seasoned automotive professional desiring to improve their management skills within the automotive business.

The Sales Department

The Sales Department usually generates the most amount of sales and gross profit in a dealership. It also generates the greatest amount of expense, such as advertising costs and floorplan interest on the vehicles in stock. It ties up a tremendous amount of capital in used vehicles and vehicle receivables. Recently, manufacturers have cut the profit margin (the difference between the suggested price and dealer invoice) and it has become commonplace for the New Car department to have a net loss. The keys are to keep the expenses low, manage the inventory, and retain the customers a dealership already has, in order to reduce advertising expense...

The dealership Sales Department is usually located in front of the dealership and is comprised of a sales lot that holds the inventory and a showfloor inside the building. This is how the nickname "front end" came about. On the financial statement there are three major areas that are considered "front end" profit.

- New Vehicle Department. This department shows the retail sales and gross profit of each vehicle model. There is a section for Car Sales and another for Truck Sales. If the dealership has multiple franchises, then the models of the "controlling" or main franchise are broken out and the other franchises are grouped in "other makes." Fleet sales are also grouped together and only broken out between Car and Trucks

- Used Vehicle Department. This section of the financial statement separates sales by used cars retailed, used trucks retailed, used cars wholesaled and used trucks wholesaled. There is a cost of sale for reconditioning, which is the cost of getting the vehicle ready for sale.

- Finance and Insurance Income. The placement of these accounts varies per franchise. Some consider them "other income" and others place them in the New and Used Vehicle department section. Regardless of their placement, there is usually not a sale and cost of sale account for this income. The only breakdown is by new and used and hopefully between income and chargebacks.

- Extended Warranty Sales. These accounts are usually in the New and Used department section and they do contain sales and cost of sales accounts. They are also named Service

Contract Sales, ESP Sales, GMPP Sales, or other names specific to the franchise.

Sales Department sales information is stored in a record called the car deal. This folder contains the legal contract copies of the sale transaction and a record of the accounting entry. The car deal folder or jacket may also be the original inventory record.

There are three different computer programs controlling the Sales Department, the Vehicle Management, Sales Management, and F&I Programs. The Vehicle Management system holds the vehicle inventory records and runs reports regarding vehicles in stock and in transit to the dealership. The Sales Management system, along with Showroom Control or Prospecting, tracks the activity of customers and salespeople. The F&I Program is often the only one of these three programs that the dealership has. It is the most important since it contains all the sales transaction data and prints the contracts and legal forms that are required for a vehicle sale.

Quick Check

☑The Sales Management programs are the most underutilized programs of the computer system (less than 30 %.) Run some of the reports and see if the sales managers knew they even existed.

DAILY OPERATIONS in the Sales Department

Steps to the Sale

Each dealership normally has a "sales system" that determines how customers are greeted and treated while shopping for a vehicle. These systems can be a "liner/closer" system where a "liner" or front line salesperson with little experience greets the customer and shows them the vehicles. As soon as the customer comes inside and starts to haggle, the closer takes over and completes the sales transaction. This closer is sometimes called an assistant sales manager. In other dealerships a straight sale process is used. The salesperson is normally more experienced and stays with the customer throughout the full transaction. Most dealerships have a "steps to the sale" process that they expect all salespeople to follow. These steps include greeting the customer, taking the customer on a demo ride, obtaining an offer from the customer, making a counter offer, closing the deal, and delivering the vehicle.

"UPS" and Prospects

A new customer is normally called a prospect or "UP". The number of prospects or "Ups" should be tracked each day and compared weekly to determine the effectiveness of advertising or follow up calls. There are various methods to track "ups" such as a multiple copy "up" card, a log, or an entry in the computer system. A typical "up" card has the customer's name, phone number, and what type of vehicle they are interested in. It is important to also obtain the ad source for the customer. Some common sources are;

- ☐ Newspaper
- ☐ TV
- ☐ Radio
- ☐ Drive by
- ☐ Referral (by another customer)
- ☐ Referral (from the service department)

Sometimes a dealership rewards existing customers who refer a new customer to the dealership. This is called a "bird dog". A "bird dog" check is issued to the referring customer and is often charged 50% to the cost of vehicle being sold and another 50% to the salesperson who sold the vehicle.

Quick Check:

☑Find out how prospects are being tracked. Is this being entered on the in-house computer system or kept on a PC? This data needs to be stored for future reports.

☑In some states "bird dog" fees are illegal because the state considers this receiving a commission without a sales license. Check your local regulations and make sure the dealer is not in violation of the law.

Demonstration Ride

After the customer has been greeted and names exchanged, they normally take a demonstration ride in the vehicle in which they are interested... Most dealerships and states require that a salesperson accompany the customer on this ride. In recent years, many salespeople have been injured by customers (and even killed) during these rides. There are a few important steps that you need to make sure a dealer is taking in order to protect their salespeople;

1. Make sure all salespeople go on a predetermined course for the demonstration ride.
2. The salesperson should inform the sales manager that he is taking a demonstration ride and leave a copy of the customer's driver's license with the manager.
3. Have a predetermined amount of time that a demonstration ride should last. The manager should investigate if the salesperson is overdue in returning the vehicle.
4. Stay in touch with local authorities about any local criminal activity.

Write-up and Buyers Order

After the prospect has decided on a vehicle, they normally make an offer. Although there is a window sticker price on the vehicle, it is common knowledge that the dealer will accept less for the vehicle. That is because there is a difference between this sticker price and the dealer's cost. With the help of their salesperson, the prospect makes an offer, usually in writing. This write up form is not the final sales invoice or buyers order, rather either a piece of scrap paper or special "4 Square" form that has boxes to take the salesperson through this haggling period. Often this write up doesn't even get saved in the car deal jacket and is discarded when the final buyer's order is prepared. The buyer's order is prepared after the customer has agreed to the final price for the vehicle.

Credit Reports

While the prospect and salesperson are haggling, there comes a point where it appears that there is going to be a sale and discussions turn toward financing. If the prospect is going to finance through one of the dealership's sources, then a credit application is completed. This credit application is taken into the sales or F&I (Finance and Insurance) office to run a credit report. To run a credit report, the first 4 or 5 lines of the credit application is entered into the in-house computer system, F&I computer system, or dedicated credit bureau terminal. Some of the most common credit bureaus are CBI, TRW, and Equifax. There is a charge of between $1.70 to $4.00 for each of these reports, depending on the agreement that the dealer has with the computer company or credit bureau. This charge is a sales expense although in some dealerships it gets charged to data processing since the charge comes on the computer company billing.

Quick Check

☑Often the computer system was set up to run credit reports from all bureaus although the dealer only uses one. They are being charged for each one. Check the setups and make sure that the default is the one bureau that the dealer used... They can manually request additional credit reports if necessary.

☑Check the billing for credit reports. Make sure these are only being run for customers and not for employees (the dealer could be liable for age discrimination if a job offer has not been made because the credit report discloses the prospective employee's age.) If the dealership is running credit reports after job offers, make sure the job application authorized the dealer to do this. Another problem is misuse. Dealership employees will run these on themselves, friends, or family members. This is normally a violation of the agreement with the credit bureau.

F&I

After the customer has selected the vehicle they want and agreed upon a price with the salesperson and sales manager, the F&I department processes the paperwork. This manager, often called a business manager or F&I manager, is also a salesperson for various products like extended warranties, credit life and disability, aftermarket items, and finance income.

Extended Warranty and Credit Life

Extended warranty is just like it sounds – it extends the warranty on the vehicle. Most new vehicle warranties are only for 3-5 years or 36,000 – 60,000 miles and after the first - third years covers only the power train. An extended warranty extends this period to five or even seven years and covers much more, like towing and rental cars. Sometimes the extended warranty company is owned by the dealership, but managed by specialty firms that invest the funds and pay the claims. Credit life is a decreasing term life policy and pays off the loan on the vehicle if the owner dies. It can be single life or joint life. There is also a disability option that will make the car payments should the customer become disabled. The accounting for these policies is as follows;

1. Credit Life - has a premium for the term of the policy. The life portion has a different commission rate for single or joint life and disability. A premium of $300.00 may be $100.00 for single life at a commission of 15% and disability of $200 at a commission of 20%. The total commission would be $55 ($15 plus $40.) Another premium of $300.00 could be $50 for joint life with a rate of 20% and disability of $250.00 with a rate of 10%. The commission would be $35 ($10 plus $25.) The commission goes to an income account and the premium less commission goes to an insurance payable account. Since it would be a problem if the dealer collected the premiums and never forwarded them to the insurance company, this account should be scheduled by customer number and aged.
2. Extended Warranty - is handled differently. The sale amount usually can be any amount determined by the F&I manager and the cost is determined by a rate chart issued by the extended warranty company (based on model, year, and coverage). The cost of the extended warranty is treated as a cost of sale. Rather than a corresponding inventory credit, in which the credit goes to a payable

account. To ease the payment of the extended warranty statement, the credit is normally scheduled by the last 8 of the VIN#, if the company bills that way. It is important that a customer name is entered into either the description field or control number 2 to make reconciliation of the schedule easier. This should also be aged to watch for older premiums that have not been forwarded.

Aftermarket Sales

These are sales of "add-on" accessories, like chrome bumpers, rust proofing, stain guard, alarms, and bed liners. In some dealerships the F&I manager sells these products and in other dealerships, a special aftermarket manager visits with the customer and tries to sell these products – usually before they go into the F&I office. The accounting for these is similar to the extended warranty sales, except the cost of sale entry usually credits the vehicle inventory account instead of a payable. The sale account is sometimes the same as the vehicle sale account and makes reporting on these sales difficult. The best method of accounting would be to have a sub account by each model for aftermarket sales. The credit would go to a scheduled due bill payable account instead of the inventory. A detail of these sub accounts would provide the accounting necessary to monitor these sales.

Quick Checks:

☑Make sure you review the schedules for all three of these items (Credit Life, Extended Warranty, and Aftermarket Sales.) These are important schedules to monitor. The company can deny claims if a policy was not submitted in time. It is easy to wind up paying for policies for customers that never bought the vehicle (unwinds or reversals.) There are many errors in costing the policies due to rate chart changes. In addition, there are refunds for cancellations that are waiting for reimbursements on the schedule.

Delivery of the Vehicle

After the customers sign all the paperwork, they are given back to the salesperson for the delivery of the vehicle. In some dealerships this procedure is handled by a special customer service manager who explains the warranty book and gives the customer a tour of the service department. The intent of this manager is to insure that the dealership has performed the steps required by the factory to receive a high customer service rating (CSI or QCP score.) Some of these steps include making sure the vehicle is clean and has a full tank of gas and having the warranty explained to them. In most states, a temporary license is affixed to the vehicle until the customer's permanent license plates arrive from the state titling office.

Reporting the Sale-License Fees

Each state's licensing procedures are different. It is important that the dealership follows the regulations regarding reporting the sale carefully. Usually, a notification is required within a few hours or days and must be mailed to the state office. Next, the titling paperwork is processed at a local office. When a customer pays for the vehicle, they also pay the license fees along with the sales tax and other charges, such as insurance and extended warranties. These license fees must be remitted to the state or county along with the proper documentation within a certain amount of time. This is the reason why the license fee account must be scheduled and aged to alert management whenever fees are not being remitted in a timely fashion. Differences between what the customer is charged and what is paid to the state must be either collected or refunded to the customer depending on state regulations.

Quick Check

☑Find out what the penalty for late submission of title documentation is in the dealership's state and what account these penalties are charged to. Often over collection of fees are credited to this account (in violation of state law) and hide the total debits. Look for excessive penalties for late submissions and find the reason for the poor paperwork flow.

Sales Taxes

Most states have a special sales tax guide for auto dealers. Review this guide to find out the taxing requirements for that state. If the dealer is not keeping the proper documentation for sales tax exempt sales, an audit by the state can be a costly expense. Typical documentation required for an "out-of-state" customer is proof that the vehicle was shipped out of state or given a "one-way transit permit" to their state. If the dealer provides a temporary license plate for the dealership's state that can often trigger a requirement to collect the sales tax. There are numerous rules for what amount is taxed and what is not. Some states let dealers exclude the amount of the trade-in from the sales tax and often different counties have different tax rates within the state. A lease has tax on the monthly payment and any capitalized cost reduction, but not on the purchase price. To attempt to reconcile the amount collected from customers for sales tax and the amount that the report requires to be paid can be difficult. The most common place to find errors is by reviewing the sales journals:

Error #1;

Account 407 Sales New Cavalier	$10,000.00-
Account 607 Cost of Sales Cavalier	9,000.00
Account 325 Sales Tax	525.00-

Sales tax rate is 5.5% - Sales tax should be $550.00. The setup in the F&I department hasn't been corrected to the new tax rate.

Error #2:

Account 407L Sales Lease Cavalier	$10,000.00-
Account 407L Sales Lease Cavalier	2,000.00-
Account 607 Cost of Sales Cavalier	9,000.00
Account 325 Sales Tax	120.00-

Sales tax rate is 6%. The sales tax report excluded the full $12,000 from taxable sales, but the $2,000 is a cap cost reduction (down payment on a lease) and is fully taxable.

Down Payments, Contracts-in-Transit and One-Pays

The are various ways that a customer can pay for a vehicle, including financing from a lender with whom the dealer has an agreement. If the lender is a credit company owned by one of the manufacturers that the dealer sells vehicles for, this is called "captive financing." This term is also used if the dealer uses that credit company as their floorplan lender. The amount financed on the contract that the customer signs is paid directly to the dealer, after the proper documentation has been received. If you look at the amount financed, it includes sales tax, license fees, credit life and extended warranty policy amounts. It deducts the down payment received and any trade-in value. Often this payment from the lender to the dealer is in the form of a draft or check that the dealer prepares or an electronic funds deposit into their checking account. Other times the lender has a special bank account that they credit contract proceeds with and the dealer can write a check from this account to pay themselves the amount owed. This amount financed is kept in an account called "contracts in transit". It helps if you have a separate schedule by each lender or the lender's name in the description on the schedule. If the customer is financing with their own bank or credit union – it is called a "one-pay." The amount of down payment on the vehicle and the total one-pay amount are kept in an account called "vehicle receivables." Common accounting errors cause the credit from a car deal to go to the contracts in transit account and debits to the vehicle receivables or vice versa. If you can run a side-by-side schedule of these two accounts, these posting errors are easy to spot and correct.

Quick Check

Some schedules allow you to have a second control number. Use the due date of the one-pay or lender name as the 2nd control number and sort the schedule based on this date or name.

Understanding Vehicle Inventory

New Vehicles

Each dealership has a franchise agreement that enables them to sell a certain make of vehicles. One dealership might have several franchises in the same company such as ABC Ford, Isuzu, and Nissan. Some factories require that each franchise be kept separate and separate financial statement produced for each franchise. Others allow you to combine the balance sheets, but require a separate sales line for each franchise type.

Allocation and Orders

Each factory has a different way in which vehicles are ordered to be built or allocated. Import manufacturers usually have a shipment of vehicles available and they divide them between dealerships based on their past sales – or allocate. Domestic manufacturers work on an order bank system. Dealers enter the vehicles that they would like built into an order bank. When the manufacturer decides to build a certain amount of a model, they then select dealer's orders in the order bank based on previous sales and preference. It is important that dealers keep track of not only what they have in stock, but also what they have in the order bank and in-transit to the dealership. This information is provided in weekly or monthly reports from the manufacturer, but the best method is to have an interfaced inventory system. An example of this is UCS and Ford or EDS and Saturn. When a dealer orders a vehicle on their in-house computer system it is added to their inventory system. As the vehicle order progresses through the manufacturer's order system from a status of being in the order bank, to being built, or in-transit, this information is updated at night to the dealership's computer system. At any time, the dealer can run a complete report of all vehicles in stock and on order from their own computer system. If the dealership's computer system is not interfaced with the manufacturers, they can still maintain the inventory on the system by entering each vehicle into the computer when it is ordered instead of when it is received. It receives an order number and the status can be updated as reports are received from the factory. The worksheet in the next section is a tool that can be used to monitor each status.

Ordering New Vehicles

There should be a written policy used by the dealership to order new-vehicle inventory. Either the dealer or general manager should be responsible for making the vehicle ordering decisions and these decisions should be based upon an inventory worksheet provided by the sales manager. The accounting office can help produce the information that is needed to make the decision. The following worksheet is a guide to ordering vehicles. It is important to know:

1. What has sold in the past by model or option

2. What is in stock by model
3. What is already on order or in-transit by model

New Vehicle Ordering Worksheet*

New Vehicle Ordering Worksheet

	Sales YTD thru 30-Aug	AVG per Month 8	60 Day Supply 2	In Stock 97	In Stock 98	Total In Stock	In Transit to Dealer	Total Available	Excess (Under) 60 Day Supply	On Order Status Order Bank	On Order Status Being Built
Model/Pkg											
1AD57 - 3KJ	19	2	5	1	1	2	16	18	13	2	10
1AD57 - 4KJ	75	9	19		7	7	3	10	(9)	4	1
1SJ37 - 3PK	12	2	3	1	1	2	1	3	0	0	
1SJ37 - 4SU	140	18	35	12	10	22	14	36	1	3	4
1SJ47 - 4SU	47	6	12	5	12	17		17	5	6	3
1SP37 - P37	45	6	11	1	3	4	8	12	1	3	
2RS27 - P40	123	15	31	14	5	19	4	23	(8)	4	6
3T437 - P40	23	3	6	3	2	5	2	7	1	2	
	484	61	121	37	41	78	48	126	5	24	24

**** Click on Contact-Free stuff form at SandraJerome.com and click on the link for Super Controller zip file of spreadsheets.***

Calculating Days Supply of New Vehicles

The following worksheet is a good tool to use to figure the days supply of new vehicles based on dollar amount. If you don't have any units over 60 days old, that isn't enough – you also should not have more than a 60 day supply.

New Vehicle Inventory

Cost of Sales New Vehicles	$ 2,352,425	MTD
Total Cost of Sales	$ 2,352,425	
Guide Months Supply*	2	
Guide New Vehicle Inventory	$ 4,704,850	
Actual New Vehicle Inventory	5,325,252	YTD
Variance (Excess)	($620,402)	

* Input 2 for 2 months (60 days)

New Vehicle Invoices

After the factory builds the vehicle for the dealer, a new vehicle invoice is produced and either mailed to the dealer or transmitted via the "mailbox." The mailbox is the nightly group of messages sent to the dealer from the factory that includes invoices, warranty credits, service bulletins, and rebate guidelines. Some computer systems can download the new vehicle invoice into the inventory system of the computer and setup a new vehicle posting record. The most common accounting posting of a new vehicle purchase is as follows:

Invoice Amount $33,550.

New Vehicle Inventory	$33,000+
Advertising expense	$50+
Holdback receivable	$500+
Floorplan payable	$33,550-

The vehicle record has two costs – accounting cost and sales cost. Hopefully these two amounts will be the same. After the new vehicle invoice is posted, the vehicle is on the new inventory schedule for $33,000 controlled by the stock number. The stock number is the sequential number assigned to the vehicle when it is received or when the invoice is received. This stock number can be sequenced many different ways;

1. Last 6 or 8 of the VIN number
2. First two digits the year- and sequential – like 99001 for the first 1999 vehicle received.
3. Sale account and sequential – like 407001 for account 407.

Used Vehicles

The used vehicle department is often the most profitable in the dealership.

Purchasing Used Vehicles

Dealerships acquire used vehicle inventory in three ways;

1. Trade-ins from other vehicle sales
2. Purchases from a local wholesaler who buys and sells vehicles between dealerships
3. Purchases from a factory sponsored or local auction
4. Purchases from customers who want to sell their vehicle without buying another

Used Vehicle Stock Numbers

The stock numbers can be assigned in many different sequences:

1. P3001 for a purchased vehicle
2. 99001A for the trade-in from new vehicle stock number 99001
3. P3001A for the trade-in from purchased vehicle P3001
4. 99001A2 for the 2nd trade-in from new vehicle stock number 99001 (deal had two trade-ins)
5. 6201 – some dealership give a new stock number to all used vehicles
6. 6201B – trade-in from stock number 6201A (trade-in from 6201)

Used Vehicle Category Analysis Worksheet

Used Vehicle Category Analysis

Category	Sales YTD thru 30-Aug	AVG per Month 8	60 Day Supply 2	In Stock Lot A	In Stock Lot B	Total In Stock	Auction Puchased in Transit	Total Available	Excess (Under) 60 Day Supply	Authorized to Buyer
A - EcCurr	19	2	5	1	1	2	16	18	13	0
B - Ec2yr	75	9	19		7	7	3	10	(9)	15
C- Ec3pl	53	7	13	1	1	2	1	3	(10)	12
D- CoCurr	140	18	35	12	10	22	14	36	1	2
E- Co2yr	47	6	12	5	12	17		17	5	6
F- Co3pl	45	6	11	1	3	4	8	12	1	3
G- MidCurr	123	15	31	14	5	19	4	23	(8)	9
H- Mid2yr	23	3	6	3	2	5	2	7	1	2
	525	66	131	37	41	78	48	126	(5)	49

The above worksheet can be used for numerous different categories such as:

1. Model type similar to the separation used by rental companies (in the example)
2. Monthly payment amount
3. Year
4. Price

This lets you know what sells at your dealership and what you already have in stock. An added benefit is the last column, which is a authorization to the used vehicle buyer to purchase a type of vehicle.

Calculating Days Supply of Used Vehicles

It is a good practice to make sure no used vehicle is over 45 or 60 days old, but you should also make sure that the dealership is not carrying more than a 45 or 60 days supply. The following worksheet uses the guide of "2" for a 60 days supply. To see how much you have over a 45 day supply – use 1.5 for the guide of months supply.

Used Vehicle Inventory

Cost of Sales UC Retail	$104,582	MTD
Cost of Sales UT Retail	$125,252	MTD
Reconditioning	$8,720	MTD
Total Cost of Sales	$238,554	
Guide Months Supply	2	Input
Guide Used Vehicle Inventory	$477,108	
Actual Used Vehicle Inventory	985,742	YTD
Variance (Excess)	($508,634)	

MANAGEMENT in the Sales Department

Employee Positions and Job Descriptions

The normal positions in the Sales Department include:

General Sales Manager
Sales Manager
Assistant Manager/Closer
Fleet Manager
Salesperson
Customer Retention Manager or BDC Manager
Greeter
Sales Secretary or Assistant
Receptionist
Customer Relations Manager

General Sales Manager

The general sales manager has the overall responsibility of both new and used car departments, their respective managers, their functions, and acts as a liaison to factory representatives. Some of the duties include development, implementation, and maintenance of a sales system and development of sales promotions. They communicate compensation plans to sales managers and help them focus on variable gross and net profit.

Sales Manager

The sales manager reports to the general sales manager or general manager. The sales manager has overall control over the sales department activities including supervision of the sales forces. One of the duties includes developing marketing strategies to increase sales and gross profit using advertising, promotions, and sales incentives. They also maintain internal control over the new and used vehicle inventories to prevent theft, vandalism, and lot damage. They should also develop a customer retention program that includes a follow-up of prospects and continuous contact with sold customers.

Assistant Manager/Closer

The assistant sales manager or "closer" reports to the sales manager. This position is common in stores that use a team sales system. They are responsible for monitoring and completing the sales of their team members. They also have the duty of providing or arranging product and paperwork training for their team.

Fleet Manager

The fleet manager reports to either the sales manager, general sales manager or sometimes directly to the general manager. In some dealerships this position is similar to a salesperson, but the fleet manager only handles commercial accounts. In other dealerships, there is a full fleet department that sometimes includes leasing and this position is equivalent to other sales managers. The broad duties include ordering and selling vehicles to fleet and commercial accounts. Since the fleet managers are often the most knowledge about ordering new vehicles, they sometimes order all the vehicles for the dealership.

Salesperson

The salesperson has the primary responsibility of following the steps that management has created, resulting in a successful vehicle sale and long-term customer retention. Their duties include taking customers on test-drives and presenting customer offers to the sales managers or assistant sales managers. After the sale is complete, they deliver the vehicle and make a follow-up call and ideally the first service appointment.

Customer Retention Manager or BDC Manager

The Customer Retention Manager is responsible for maintaining high customer retention and customer satisfaction, as well as monitoring customer complaints. Along with these duties they also provide reports and listings to salespeople and management of customers who are not coming back in for service, customers who may be ready to purchase another vehicle, and current factory customer satisfaction scores.

Greeter

The greeter is a new position in dealerships. The greeter meets customers coming onto the premises and finds out the nature of their visit. If they are looking for a vehicle to purchase, the greeter gets basic information and enters that into a log or computer. They then route the customer to the next available salesperson, as well as informing management. If the customer is at the dealership for service or parts, the greeter provides directions and assistance to those departments.

Sales Secretary or Assistant

The sales secretary or assistant provides clerical and administrative help to the sales managers and sales force. This assistance includes maintaining inventory records, issuing purchase orders, tracking phone calls, and maintaining logs and reports.

Receptionist

In some dealerships the receptionist works for the accounting office and in others, they report to the sales managers since their work schedule closely follows the hours that the sales department is open. Their duties include keeping track of the calls that the dealership receives each day, especially sales calls and which salesperson assisted the customer. They are also responsible for knowing the sales team schedule and keeping track of employees who are absent, out of the office, or with customers.

Developing Long-Term Employees

One of the biggest personnel challenges that faces a sales or general sales manager is how to keep good employees during the "slow" times. Most dealerships have high and low cycles throughout the year and from year-to-year. The typical salesperson is already trained in the dealership's procedures and the product line, so reducing turnover is important. Productivity is measured by sales and gross profit per salesperson and when a dealership experiences a downturn, the sales department appears to be over staffed and nonproductive.

Productivity should not be confused with long hours, overtime and job related stress. These attributes may be present when a person assumes a new position but should, in this instance, be only of short-term duration. The goal of all productive people is to find an easier and more effective method to accomplish certain tasks. The terms of each task should be reviewed and fully comprehended before changes are made. The process of change is an absolute, given our business environment. When analyzing sales department personnel for productivity, it may help to use some of the benchmarks below as a guide. One of the best solutions for expense control with regards to personnel management is to create pay plans that reward employees for productivity and customer retention.

Developing a Customer Retention Program

One of the most effective methods for increasing customer retention is to make the first service appointment for the customer at the time of delivery of the vehicle. This appointment is made by the salesperson that asks the customer how many miles they expect to travel per month and based on the response, sets a tentative future appointment date. Prior to this date, the salesperson reminds the customer either by telephone or by postcard. The whole project has been controlled by using sophisticated software using a timekeeper function to remind salespeople of when to make the customer contact. This procedure has led to a dramatic increase in the first service appointment show rate. The cost of retaining a customer, rather than going out and trying to acquire a new one, is less. Recent studies have shown that a repeat customer means additional incremental income to the dealership - income that is above and beyond the average gross profit earned on selling a vehicle - in excess of $1,200! Therefore, it pays for the dealership to create a strategy of improving its customer retention rate.

Which Salesperson gets Orphaned Customers?

Do not assign orphan customers arbitrarily to a current salesperson – because if that salesperson leaves and you assign it to another salesperson that then also leaves – you'll lose track of who was the original salesperson. Instead, provide the contract sheets with the former salesperson's name on it to other salespeople. When they have made contact wit the customer – via a phone call or email – then reassign the individual customer to that salesperson.

Encouraging Employee Honesty

Internal controls (discussed in a later section) are an essential management function to protect against theft. Another method is to encourage employee honesty through example. It is difficult to expect sales employees to be honest about the supplies when the dealer, other managers, or even the sales manager take gas, T-shirts, and office supplies out of the sales department.

Providing Sales Training

Sales department employees should attend regular product and customer service training seminars. These employees spend more time in contact with customers than most other employees. It must be understood that customers are not just the people who fill out surveys, but they are also the other employees, the service department, and factory personnel. It is important to establish a good relationship with the other departments. The sales department depends on repeat sales for future customers.

Quick Check

☑Make sure you review a factory survey report that details responses by salesperson. Make sure sales are being reported with the correct social security numbers to make this report valuable.

Quick Check

☑The accounting office should maintain a training record for each employee that shows which training seminars employees have attended (especially sexual harassment seminars.)

It is important to study the process and flow of paperwork through the Sales Department. After the addition of computers, many manual logs remained and were used parallel to the computerized system. In an effort to reduce expenses and increase productivity, each of the different flows must be analyzed. Steps that are redundant and do not contribute to either gross profit or internal control should be eliminated. Other processes need to be combined and performed by the same person for efficiency. Anytime a form can be eliminated it should be. Study the car deals and see if anything that is "handwritten" can be programmed by the computer to print. Look at all logs and see if there is a computer generated report or display that can replace the log. Each step in the paperflow process can slow down the processing of the car deal and delay payment to the dealer by the bank and delay paying off the floorplan balance. Other flows to study are:

1. The flow of new vehicle invoices to stock cards. Most of this has been automated by the computer system, but is still being done manually.
2. The flow of commission vouchers to commission summaries. This is tracked by the computer and manual vouchers can be printed in F&I and the summary replaced by a computer report.
3. The flow of purchase orders for added equipment. Again this has been automated and often a duplicate system is in place.

Breakeven Analysis

It is important that the Sales manager know what it takes to breakeven each month. The chart below is an example of a break-even analysis. The breakeven analysis tells you how many vehicles the dealership has to sell to breakeven. It takes into account the variable expenses per unit and F&I income. The first step is to enter the parts and service gross profit average for the month. Next, find the non-variable expenses (total expenses less other income and variable expenses.) Third, enter the gross profit average per unit, net F&I income per unit and a variable expense percentage.

If done on a spreadsheet:

Total expense with variable: (Non-variable expenses – parts and service gross profit)+ variable cost per unit X breakeven point in units
Net profit per unit: Gross profit per unit + F&I income per unit – variable cost per unit
Breakeven point: (Non-variable expenses – parts and service gross profit)/(Net profit per unit/(Gross profit per unit+F&I income per unit))
Breakeven point in Units: Breakeven Point/(Gross profit per unit+F&I income per unit)

ABC MOTORS
Breakeven Analysis

Parts and Service Gross Profit	$125,000
Non-variable expenses	$350,000
Gross Profit per unit	$1,200
F & I Income per unit	$300
Variable Cost Per Unit	$276
Variable percent	23.00%

Total expenses with variable	$275,735
Net profit per unit	$1,224
Breakeven Point	$275,735
Breakeven Point in Units	184

Forecasted Units	200
Forecast Gross Profit	$300,000
Forecasted Total Expenses less P&S Gross	$280,200
Forecasted Profit	$19,800

Internal Control - Expenses

Advertising Expense

The advertising expense should be accrued at the beginning of the month based on the budget provided by the sales managers or general sales manager. A purchase order for each vendor should be created and it should be made certain that advertising invoices do not exceed the amount of the purchase order.

ADVERTISING BUDGET

Month:	July		Department:	
TV:	Channel	$		
	KONP	$ 4,323.00		
	KOMO	$ 9,000.00		
			Total:	$ 13,323.00
Radio:	Station	$		
	KVVP	$ 3,200.00		
			Total:	$ 3,200.00
Newsprint	Newspaper	$		
	Star Report	$ 12,352.00		
	Central News	$ 32,525.00		
			Total:	$ 44,877.00
Yellow Pages		$		
	July	$ 421.00		
			Total:	$ 421.00
Direct Mail		$		
	CVC Inc.	$ 1,300.00		
			Total:	$ 1,300.00
Others	Describe	$		
	GM Coop	$ 1,358.00		
	State fund	$ 500.00		
			Total:	$ 1,858.00
Manager's Signature:			Overall Total:	$ 64,979.00

Advertising Effectiveness Analysis

There is an easy, two-part process to track advertising expense and it's effectiveness. With the data captured and stored in your computer database or an unused F&I field in other systems), run a monthly comparative analysis report of advertising spending compared to the responsiveness of your customers.

	Advertising effectiveniss analysis							
ACTUAL EXPENSE				SALES SURVEY			VARIANCE	
				Sold			Effective	Discrepancy
Media	Amount	% of Total		Units	% of Total		Amount	Over/(Under)
TV	6,300	42%		21	40%		6,000	300
Radio	1,350	9%		17	32%		4,800	(3,450)
Newsprint	5,700	38%		12	23%		3,450	2,250
Other	1,650	11%		3	5%		750	900
Total	15,000	100%		53	100%		15000	-
			Cost Per Unit Sold	$ 283				

Be sure to match the expense period to the customer traffic period that the advertising represents. For example, if the advertising is for the last weekend of the month the dealership usually will not get the invoice until the following accounting period. Be sure to accrue for the expense in the same month that the advertising occurred. Conversely, when advertising is paid in advance and contracted for a period that spans more than one month, prepaying the expense is appropriate.

Add'l .05% if interest expense is less than 10% of total gross
12% of gross less any vehicle receivables due of 60 days
20% override if no new units over 60 days old

1. Reduce the amount of time it takes to get the vehicle ready to sell. Plan ahead:

   At least 3 days before the vehicle arrives:

   After you receive the Invoice:

   - Office assigns a stock number and updates computer
   - Copy of the invoice goes to sales department
   - Copy of the invoice goes to get ready to check the equipment when it arrives
   - Key tags are made
   - Dealer sticker typed
   - Jacket is made
   - Internal is opened in service and SCHEDULED for arrival date

   Arrival Date

   - Get ready checks for equipment and damage
   - If there is either, copy of invoice and shipper to warranty administrator
   - Vehicle is parked in service - preferably the tech's stall - we are a waiter!
   - Get ready exchanges manual and extra keys for office's typed dealer sticker and key tags
   - When PDI (pre-delivery inspection) is done then to get ready for clean-up
   - Park in a special area – all salespeople view new arrival!

   What can this save?

   $3,000,000 x 8.5% /365 = $698 per day

   $698 per day

x 5 days
$3,490 per month
$41,880 per year

Gas Expense

The following is a list of some of the controls that should be in place for gas expense in the sales department:

1. There should be a purchase order or authorization form that allows gas to be put in vehicles in inventory for demonstration rides. These forms should indicate the customer and stock number. This information should be compared to the "up" log (listing of prospective customers.)
2. There should be some reconciliation of the gas for new vehicle deliveries. This would include either scheduling the account against credits received from the factory for filling the tanks or a pre-made purchase order that is kept in the warranty book.
3. Employees who drive demonstrators should not use the gas tanks to fill their vehicles. They should be given a flat amount each month, for example $50 to reimburse themselves for gas.

Forecasting and Planning Benchmarks-Sales Department

There are certain productivity goals that are regarded as benchmarks in the automotive industry. They are national, but in almost all instances, will apply to most dealerships. In computing pay plans certain considerations should be given to using acceptable pay for acceptable productivity.

1. 25-30% net profit to gross
2. 40% Customer retention rate
3. 3-4% Net profit to gross sales
4. Gross profit per employee - $6,000
5. Advertising as % of N/U Gross – 9%
6. Unit Projections - Better than market share
7. Leases - 25%
8. Finance penetration - 70%
9. Warranty penetration - 50%
10. Inventory 45 days supply
11. Sales and management compensation cannot exceed 30% of vehicle gross.
12. F&I compensation should not exceed 10% of F&I gross
13. Policy 2% of Gross
14. Floorplan 2-7% of Gross
15. Salesmen to managers 5-1

The Service Department

The Service Department is the busiest and most problem prone of the gross profit generating departments of a dealership. It usually has the greatest amount of paperwork, employees, and customer contact. If it is managed well, the Service Department can contribute an excellent net profit. It is also very common for the service department to have a net loss. The key is to keep the expenses low, manage the labor "inventory", and keep the customers happy.

The dealership Service Department is usually located in the back area of the dealership and often attached to the parts department. This is how the nickname "back end" came about. It is attached to the parts department because the service department is often its biggest customer. On some dealership financial statements the parts sold on repair orders is a service department gross profit (GM), and on others the profit is kept in the parts department (Ford.) To determine which the case is, look at the service and parts sections of the financial statement. The following is an explanation of all the service sales and gross profit accounts:

- Customer Labor (Cars and Trucks) is repair labor done for customers that is not under warranty, but paid for either by the customer, insurance, or an extended warranty company (except for Ford that shows the Ford Extended Warranty under "ESP Repair Shop Labor".) The gross profit percent is usually 60-75%.

- Warranty Claim Labor is repair labor that is being reimbursed by the manufacturer of the vehicle. The gross profit percent is usually 60-75%.

- Internal Labor is created when the dealership adds accessories to vehicles like cruise controls, and reconditions used vehicles. The gross profit percentage is between 50-75% .

- Repair order parts sales are sales of parts on repair orders in the service or body shop departments. The gross profit percentage is between 35-50%. Note that on a GM Statement this is above line 32 and included in "Total Mechanical". There is an internal transfer of profit from the service department to the parts department to compensate the parts department for these sales. On a Ford Statement these are in the left column and included in the Parts Department Adjusted Selling Gross.

Repair order sales are created on a repair order or work order. Most computer systems control these sales internally, meaning that the labor and part transaction is stored in the computer system. The system generates the next repair order number when the repair order is started. The system relieves the parts inventory as soon as the quantity is entered, regardless of whether the sale is completed or not. As soon as labor is entered on a line, the amount due the technician is created. Both the labor and part remain as "work in process" on an open repair order listing. To determine if the dealership uses this system, look at a repair order. If the parts and labor are written on the repair order by hand (manual system, not computer generated) then a hand written open repair order list must be compiled. The amount of labor paid to the technician is shown on the back of the hard copy. We will explain this further during dispatching and booking.

There are two different computer programs controlling the Service Department, the service invoicing system and the service dispatching. The invoicing system is normally interfaced to accounting and determines which labor lines go to which account numbers. It determines the sales rate, the labor cost, the parts markup, and final billing accounts. The other system, dispatching (or electronic technicians) determines when the service work will be done and by who. It also coordinates the service appointments for the day, week, or month. The skills of the technicians are kept by this system and used to decide if a technician has the qualifications to do the work and the available time.

Quick Check

☑Make sure the service Department opens a repair order for any work done in the shop. You can check this by looking at the technician flags and verifying the repair order numbers.

☑un the open repair order listing to make sure that "work in process" is not excessive or overage.

DAILY OPERATIONS in the Service Department

Service Write Up

If a customer has made an appointment for service work and the dealership is highly computerized, then the service write up can be as easy as getting the customer's signature on a preprinted repair order. Unfortunately, many customers still arrive at a dealership without an appointment or advisors don't use the computer fully. What normally happens is the customer pulls up to the service drive and waits for a service advisor to come up to them. The advisor has a write up sheet on a clipboard and proceeds to get the customer's information and the vehicle information. He then writes down the customer's complaints, asking questions about the previous difficulties or repairs. Depending on the state requirements, the customer either signs the write up sheet, or waits while the advisor prepares a computerized work order for the customer to sign. The customer is either a "waiter" that will stay at the dealership until the vehicle is done, or given a promised time. The customer is often given a free rental vehicle or courtesy ride to work or home. There is usually a customer lounge for the waiters that has TV and coffee.

Quick Check:

☑Suggest that the appointment system be used and appointments are staggered during the day to prevent the 7:30 a.m. bottleneck in the service drive.

☑Make sure that estimates are being given to customer if it is required by the state and that the labor rate is posted or disclosed on the repair order if also required.

History Inquiry

Hopefully sometime before the customer arrives, the advisor checks the history of the vehicle. There are three vehicle history files. The first one is the hard copy file. These are the physical hard copies of all previous repair orders normally filed by the last 6 or 8 of the VIN#. These hard copies contain the original technician notes and actual time spent working on a vehicle. The second history file is stored on the computer system and is closer to the information on the front of a repair order. The third history file is accessed online to the manufacturer. For a Chevrolet dealer this is called VISS and for a Ford dealer it is an OASIS. The dealership's DCS computer calls the factory computer and requests a record of all warranty work done to a vehicle. This history record also contains the work done by other dealers, the delivery date, the selling dealer, any open recalls that need to be done, and sometimes extended warranty information (Ford -ESP.)

Quick Check

☑Make sure the dealership is running an online history inquiry for all inline vehicles (Fords for Ford dealers, Chevrolets for Chevrolet dealers.) The factory wants recalls done and judges a dealer on their ability to perform these. It is also a missed profit opportunity for the recall and a protection against performing the same warranty work twice and not getting paid for it.

Recommended Service

Service Merchandising consists of recommending service to be done based on industry and manufacturer guidelines for a certain model. The best time to make these recommendations is during service write up, but often the step is either forgotten or time does not allow. The easiest solution is for the dealership to develop "menu" items that can be kept on a flyer. When the advisor notices that a vehicle has 33,000 miles on it, he can suggest the 30,000-Mile Menu item and give an estimate from the flyer. If the dealership has merchandising software, this recommended service should print on the work order based on the setups done by the dealership and the vehicle's history. You can see the importance of using the appointment system mentioned above. If the customer makes an appointment, the advisor or service receptionist can ask for the mileage and the recommended service can be suggested and scheduled before the visit.

Making Appointments for Service

The objective in any well run shop is to have over 80% of the customers for the day be from appointments. Having a customer make an appointment enables better dispatching of the work, more preparation done in advance such as history inquires and preprinted repair orders, and quicker service due to staggering the flow into the service drive. The key to getting appointments is having a dealership employee call the customers on a regular basis. This can be done based on follow up reports generated by the in-house computer or by an outside service that specializes in this type of follow up. To monitor if appointments are being made, ask for a copy of the appointments for the day first thing in the morning. Match this list with a report of repair orders opened and completed at the end of the day and you'll find the percentage.

Quick Check

☑To improve the amount of appointments, have the sales person make the first service appointment at the time of the sale.

Dispatching

After the customer has been written up, the information is entered either into the computer system or onto the repair order. In a computerized system, the advisor must enter the priority level for the customer. This tells the dispatching system if the customer is a "waiter" and must be done immediately. The priority level could also indicate that the customer bought his car from another dealership and should go in line after the dealership's own customers. Next the advisor enters a sale type; customer, warranty, or internal. Sales types determine the pricing level for labor and parts and accounting information. The advisor then determines a technician skill level for the job and sometimes enters in the labor operation codes. The labor operation codes determine how many hours it will take to complete the job and often the skill required. If the advisor is busy, the labor operation code may not be entered until the billing or booking process that is discussed later. After the information is entered, a "hard copy" is printed and taken to the dispatch office. If the dealership is on manual repair orders, the same is done except the information is hand written on all copies of the repair order including the hard copy. Sometimes both the computer printed form and manual repair orders have an extra copy for the advisor that is pulled at this time to help them track open repair orders. If this is the case, then the extra copy is normally logged into an advisor's log for future follow up.

When the dispatching office receives the hard copies, they finish entering any necessary information into the computer and lets the computer dispatch the work to the best available technician or team. The computer uses the available skill levels of the technicians that are not working to determine who should get the next job. If the dispatcher is manually dispatching, then he logs the repair order onto a large dispatch sheet. He usually makes a determination based on his knowledge of who will get this job and puts it into a sorter by technician or job type (heavy, light, or specialized.) Once the technician receives the hard copy, he normally punches the hard copy into a time clock and also punches his flag sheet into the time clock. The start time for the job is recorded on both. Throughout the day, the dispatcher monitors either the log or computer screen of jobs that are working and complete. When a job is completed, the technician punches the stop time on the hard copy and his flag sheets and brings the hard copy back to the dispatcher to get another job. Sometime during the process, the dispatcher, advisor, technician, or booker determines how many hours the technician will get paid for the job and enters it into the computer and completes the flag sheets.

Proper dispatching is the key to technician productivity. If technician productivity is low, the dispatch office may be the source of the problem. Observe the method that the dealership uses to dispatch and ask the following questions:

- Does the dealership have the dispatching (scheduling) option on their computer system?
- If so, does the dispatcher let the computer dispatch the next ticket, or does he manually override the system? The reasons why he may override the system could be that the skill levels and priority levels are not set up correctly and the system must be overridden. Another reason why he may override the system is favoritism to certain technicians. Observe if any technicians that refuse the jobs that the computer assigns them.
- If the dealership uses a manual system, look at the dispatch sheet. Is there an indication of which technicians are doing the jobs that are logged?
- Are completed jobs updated on the log?
- Is there a holding area for hard copies that sorts them by either technician or job type?
- Does the dispatcher keep track of labor inventory? Does he know by job type or technician how many hours are left to sell for the day?

Electronic Repair Order

In a highly computerized dealership, the dispatching function may be done by the advisors and technicians. This system is called ERO or ESI. Technicians have their own terminals and they receive their next jobs via the terminal. They log on and the computer system determines the next job for them to complete. They accept the job and the system keeps track of the time. When they have completed the job, they go back to the terminal and update the repair order with any notes and then receive another job. Hopefully if the dealership has this system they are using it as intended. There should be no hard copies and no technician flag sheets. All the technician flags are kept by the computer system and downloaded to the payroll system.

Booking Labor

At some point after the job is complete, someone needs to determine how many hours the technician will get paid for a job. Unfortunately, this job is often done by too many people in the dealership. In an ideal situation, there is a booker that determines the labor operation code and flags the technicians. Both the technician's flag sheets, the computer record, and the hard copy are updated at the same time. The booker looks at the technician notes and determines what was done. From this information, he looks either at a labor time guide manual, a microfiche record, or interfaced computer labor time guide. The proper labor operation code and the time allowed for the operation is obtained. If the labor operation provides 2.3 hours, then the booker writes the repair order number on the technician's flag sheet next to the punched start and stop time, and writes 2.3 next to the RO number. The top copy of the flag sheet is perforated in ½ inch increments and can be licked and affixed to the back of the hard copy of the repair order. When a full sheet is completed or at the end of the day, the sheet is totaled and a copy given to the technician and the payroll department.

Parts on Repair Orders

In many cases the technician will need parts to complete the job. The technician completes a parts requisition form and brings this to the back parts counter. The parts counter person does an inquiry for the part number and either gets the part off the shelf or tries to locate the part from another dealership. If this is unsuccessful, then he completes a special order request and the part is ordered. He informs the technician and dispatcher of the order and the repair order is put on "parts hold" status.

If he finds the part, the parts counter person enters the part number and quantity on the repair order and gives the part to the technician. If the damaged part has a core value or must be returned to the factory as required by the manufacturer, then the parts counter person is responsible for getting the damaged part back from the technician.

Quick Check

☑A few things can be done to improve technician productivity during the parts requisition process. The first suggestion is to have someone from the parts department deliver the parts to the technicians. Another suggestion is to create pass through cubbyholes under the parts counter, so the technician can pick up the parts instead of waiting for the back counter person to hand them to him.

Cashiering

Cashiering is the final step in completing a repair order. After the labor and parts are entered on the repair order, it is given to the cashier to close. The cashier needs to know the payment information: cash, check, or charge. The customer usually provides this information to the cashier. This payment information is entered as a payment code into the computer. Hopefully all the labor and parts pricing has been done prior to the repair order coming to the cashier. Sometimes the cashier needs to enter the miscellaneous charges such as hazardous waste and shop supplies. The tax is calculated, then the total computed. After the customer provides the payment information, the cashier either notes this on the repair order or enters it into the computer. A final invoice copy is generated and a copy given to the customer. With computerized repair orders, these copies of the repair order are called repair order invoices and the copies that the service advisor, dispatcher, technician, and parts department work with are called work orders. This means that there are sometimes 7-10 copies of the repair order generated by the computer system for each customer. Hopefully the dealership puts customer pay, warranty, and internal repairs all on the same invoice. This streamlines the paperwork tremendously. Unfortunately, some computer systems require that all pay types be closed at once to complete the repair order. If a customer comes in and has warranty work done and some quick service like a lube, oil, and filter (LOF), the repair order cannot be closed until the warranty administrator finalizes the warranty portion of the repair order. This creates a timing problem with cash sales. The customer has paid for the repair order, but the accounting copies will not go through until a few days later when the warranty is closed. To correct this problem, some dealers put warranty and internal repairs on separate repair orders from cash sales. The best solution is to require that the warranty portion be closed immediately. This requires that the service advisor, technicians, dispatcher, and booker be careful to complete all the information that is required for a warranty such as trouble and failure codes.

Another problem arises when the cashier closes a ticket but does not receive the payment. On some computer systems the cashier must close the ticket to get it ready for the customer to come in and pay. If the customer does not pick the vehicle up for a few days, the accounting entries show a closed repair order, but no cash payment. If there is a problem with cash sales created by these examples, then it may be easier to schedule cash sales for better control. Any open repair orders or credits can be investigated when they become overaged.

The cashier starts the day with a set amount of cash in her cash drawer. During the day, she collects funds for repair orders and sometimes parts tickets. It is best that she either keep a log of all her collections and balance that to the cash drawer or use a cash register tape. At the end of the day, she leaves the set amount in her cash drawer and turns in all the collected funds and documentation to the office. The cashier should not do the bank deposit herself. If the dealership is computerized, the office should receive a report that indicates which invoices tickets have been charged to cash clearing. This total should agree with the documentation provided by the cashier. Two deposits are normally done; one for the parts and service cash sales and another for the miscellaneous funds that the cashier collected (down payments, car payments, license fees, etc..)

Charges are usually either fleet accounts or payments due from extended warranty companies. Most dealerships do not have a good method to bill companies for the work that has been done. As a suggestion, the extended warranty billing should be on a separate accounts receivable schedule and given to the warranty administrator for billing and collections. The cashier should have a written policy for billing customers and all charges must be approved by the accounting office. Although this is usually controlled in the parts department, the same control is often lacking in service. Repair orders are left open and in the "bucket" until a customer pays or someone tells the cashier to write off the ticket. A daily listing should be made of the completed repair orders in the "bucket" and someone should note if the vehicle is still on the premises or if it has left the dealership. If it has left the dealership, then it has become a charge by the nature of what happened. A special schedule should be made for these "unauthorized charges" and the customers billed anyway. It is better to complete them to accounting where they can be collected and monitored instead of leaving them in the cashier's office as an open repair order.

Quick Check

☑Make sure that cash sales clearing has either a zero balance every day or is scheduled. If there is a problem with the daily balancing of cash sales, suggest that no checks be cashed from the cashier drawer until the problem is solved. This makes reconciliation easier.

Understanding Service Inventory

The service department is the only department that doesn't have a physical inventory. The new car department has new cars and the parts department has parts, but the service department only has labor to sell. The problem with a labor inventory is that it spoils easily. When an hour passes, the hour is gone and you can never get that hour back. Unfortunately in the car business, the service department does not pay enough attention to managing their inventory of time. They may know how many repair orders were written yesterday, but can they determine how much more labor could have been sold based on 150% productivity? An hour of non-productive time can never be recovered. Working faster or beating the flat rate to recover wasted time does not solve the problem because now there is one less hour available to work faster at.

Measuring Technician Productivity

There are three hourly amounts that are monitored for technicians. The first is the flat rate time. The flat rate time is the time that is authorized to do a labor operation. These times are determined by industry manuals and the factory warranty policy. Each manufacturer issues Labor Time Guides that state how much they will pay for each labor operation. The total flat rate hours are either obtained from their daily time tickets (flag sheets), or a computer generated report.

The second time that is measured is available hours. This is the total hours that a technician was at the dealership for the day ready to perform service work. The total available hours are obtained from a weekly time card.

The third time that is monitored is actual time on a job. This is the punch on and punch off time for a particular job. The total actual hours are either obtained from their daily time tickets (flag sheets), or a computer generated report.

Sometimes a fourth time is analyzed, which is opportunity hours. These are the hours that the technician had available to him to work. An example of these four hours are:

174 Hours Total hours that the shop was opened - opportunity
166 Hours Since the technician was sick one day he had 166 available hours.
189 Hours Flat rate hours - the amount he will be paid for based on the flat rate guide.
150 Hours Total actual hours that it took the technician to do the 189 flat rate hours.

Based on these amounts, we can determine a few productivity ratios:

Technician Productivity =

$$\frac{\textbf{Total Flat Rate Hours}}{\textbf{Total Available Hours}}$$

$$\frac{\textbf{189}}{\textbf{166}} = \textbf{114\% Productivity}$$

Technician Efficiency =

Total Flat Rate Hours

Total Actual Hours
189

150 = 126% Technician Efficiency

Technician Opportunity =

Total available hours

Total hours shop was open
166

174 = 95% Opportunity

Shop Technician Utilization =

Total actual hours

Total available hours
150

166 = 90% Utilization

The first two measurements, Productivity and Efficiency are the most common ratios that are monitored. Often the dealership mistakes one measurement for the other or uses total shop hours for available hours and actual hours. This is because it is difficult to track available hours and actual hours. Normally the first measurement, Technician Productivity will be enough of a guide to discover if there is a problem. If it is necessary to pinpoint the problem, rather than blaming the technicians, the other 3 measurements should be taken. It would be terrible to blame a technician for low productivity when you did a further analysis and found that it was actually poor utilization of the technician's available time. It is possible, as in the above example for the technician to take only 150 hours to do 189 hours, but the dealership wants over 125% productivity. The technician is actually 126% efficient with the work he is given, but possibly just doesn't get enough work. By understanding all four measurements, you can use the one that will find the true cause of the problem.

Technician Productivity

Flat rate hours/ Available hours

This is the standard for measuring a technician. If a technician is there for 166 hours and does 189 hours of flat rate, then he has 114% productivity. Most dealers expect between 125% and 150% productivity, so some of the problem may be due to:

- Lack of proper tools
- Insufficient training
- Poor dispatching of jobs
- Delays in getting parts
- Too many technicians or not enough work

Technician Efficiency

Flat rate hours/ Actual hours

Upon further investigation, the efficiency can be determined. In the example above, the technician only took 150 hours to do the 189 hours of flat rate work. This was actually very efficient (126%) and would mean that the actual problem was something else. If he had not been efficient, then the cause might have been:

- Lack of proper tools
- Insufficient training
- Delays in getting parts

Technician Opportunity

Technician's total available hours/Total hours shop was open

If the shop is open 174 hours, but the technician was only there for 166 hours, then some opportunity for work to be performed was lost. This may not be entirely the fault of the technician; often training and scheduling could be the problem. The time card needs to be reviewed to find if the nature of the absences was due to:

- Chronic illness
- Excessive training
- Lack of motivation
- Personal problems

Shop Technician Utilization

Total actual hours/Total available hours

If the technician only spends 150 hours on actual jobs, but was available for work for 166 hours, the low percentage would normally be a management problem. The technician was there and available to do work, but not enough work was assigned. The problem could be;

- Failure of service advisors to sell available hours
- Poor dispatching
- Too many technicians
- Not enough business, poor follow up or promotions
- Time wasters

Unapplied Time

When the service manager or dispatcher decides to pay a technician for work that isn't charged out on a repair order, this is unapplied time. Some examples of unapplied time are:

- Cleaning the shop or helping other technicians
- Paying a technician for more hours than the flat rate guide or warranty manual allows - usually done if the job was difficult or the proper tools were not available.
- Reconciling entry when adjusting the work in process account. (See later section on Work in Process.)

The unapplied time account is usually a cost of sale account number that appears on the financial statement as "ADJ - Cost of Sales - Labor". If there is a large debit balance in this account, then you need to find out why. If the debits are coming from the payroll sources, review the daily time tickets to find the reason. There is a chance that training or sick time is being charged to this account. If so, suggest that they use the expense accounts for this purpose. If they are constantly paying the technicians for more than the guides allow, make a suggestion to management to minimize this practice. Any large debit adjustment from the general journal must be researched. See the comments in the Work In Process section later. If there are large credit balances in the account, then the service department is probably using an estimated cost for the technician time instead of actual. This practice should be discouraged since the actual cost is easily obtained. For example, if the dealership has 4 technicians making, $14 hour, $13 hour, $10 hour, and $8.00 hour respectfully, then the dealership may use $13 an hour for costing purposes. Since the actual cost is obviously going to be less, there will be a large credit balance at the end of each month. It is not that difficult to use the actual cost. The cashier merely looks up the rate for the technician number on the repair order and figures the cost correctly. If a dealership has computerized dispatching or ERO, there is no problem, the computer does actual costing if the technician setups are correct. If they are on manual repair orders, they often set up the computer journal to automatically cost based on either a flat rate or a percentage of labor. Although this may sound simple, it would be much more accurate to override the automatic costing with actual costs. Without this procedure, any of the other unapplied labor problems will be difficult to detect since the credits will overshadow the debits.

Comebacks

When a customer brings the vehicle back for the same problem, this is called a "comeback repair". Dealerships handle comebacks in different ways. Some service managers require that the technician that fixed the vehicle originally fix it again for free. Others will assign a new technician to the job and "back flag" the original technician for the time. The most common way is to assign a new technician to the job and expense the cost to policy adjustments. The reasoning behind this is to not spend a lot of wasted time finding blame, but rather to get the customer's vehicle fixed by the best technician and have the expense paid by the service department. If a certain technician must have a large amount of his repairs redone, then he is either coached or terminated. Hopefully there is a method to track how many comebacks were caused by each technician. Many service managers track this "in their head", but there are better methods. The easiest method is to schedule policy expense as a balance forward schedule by the technician number. It is important that you use the technician number of the original repair and not the technician number of the person that performed the correction repairs. Some of the computer systems have a comeback report by technician, but often it incorrectly gives the comeback repairs by the correcting technician also.

Training Time

The dealership pays for the time that a technician spends at school and performing self study courses at the dealership. This time should be noted and approved on the time card and charged to the training expense account number and not to unapplied time. When analyzing this account there are normally two charges, one for the technician's time and another for the cost of schools and training materials. Make sure that service manager or dispatcher notes on the time card which courses were completed during the training time. Often a dispatcher will try to be a "buddy" to the technicians and flag them for training time if they didn't have any other work to do. The most common indication of this is when a technician has 6 hours flagged for a day and another 2 as training time. To verify training time, ask for a list of completed courses for all the technicians and match this list to the training time on the time cards. If there is time expensed for courses not listed, there may be a problem.

Sublet Sales

A sublet sale is when the repair work or addition of accessories for a vehicle will be done by an outside repair shop or vendor. An example of this is wheel balancing. Often the dealership will not have the equipment to perform this type of work and they will send the work to a local shop. The service advisor sells the repair work and often calls the local shop to make an appointment. The work is performed, then the local shop takes the invoice to the Parts Department to get a purchase order. Unfortunately, the control over sublet repairs is handled in many dealerships by the Parts Department; and they don't get paid on sublet repair profit. The reasoning behind the Parts Department handling sublet is that they have the purchase order book and can authorize payment for sublet repairs. Unfortunately, they usually don't know what really needed to be done to the vehicle or when it needed to be completed. They often don't "mark up" the invoice when they transfer it to the repair order and the profit is lost. If the dealership wants to make money on customer pay sublet, they'll need to transfer control from the Parts Department to the service advisors and pay them on it. This means either giving them a purchase order book or giving the task to the service booker. If they have electronic purchase orders, the process is simplified. If they still want to leave the control in the Parts Department, they should consider paying them on customer and internal sublet repairs. Warranty sublet repairs are normally required by the factory to be sold at cost. Rather than customize their pricing matrixes, many dealers have cost = sale for the default for all sublet repairs and are only sometimes manually marking up customer and internal sublet repairs. They can get their computer system to mark up sublet repairs automatically. For Reynolds & Reynolds users, the sublet markup is under the Make Billing Guide (3624). For CDK users, sublet is controlled by job OP UOC. There is a percentage markup field for each labor type. I would suggest that they start out with customer sublet at 40%, internal sublet at 25%. Should you mark up internal sublet repair? The best way to get the new and used departments treated like customers is to charge the same for internal sublet repair as they do customer sublet repair and pay the service advisors on internal sublet repair. If they make a sublet repair profit of $100.00 in the shop, they'll normally pay an advisor and manager a commission of between $7.00 and $9.00. If that same profit goes back to the sales department, the commission is usually between $25.00 and $40.00. If the sales managers refuse this mark up, they might want to try to get a better price from their suppliers for internal sublet. To find out if the dealership is marking up sublet repairs, look at the gross profit for sublet repairs on the financial statement. If there is little or none there, you might want to investigate further. Unfortunately, most financial statements group warranty, customer, and internal sublet together. They'll need to set up some sub accounts for each type of sublet sale and cost of sale; customer sublet repair, warranty sublet repair, and internal sublet repair. Then have them put these new accounts on the DOC to get the gross profit percentages by type. Now they can monitor if they are making money on sublet repairs.

MANAGEMENT in the Service Department

Employee Positions and Job Descriptions

The normal positions in the Service Department include:

Service Manager
Assistant Service Manager
Service Advisor
Dispatcher
Shop Foreman
Technician
Booker or Biller
Cashier
Warranty Administrator
Customer Relations Manager

Service Manager

The service manager reports to the general manager or the parts and service director. This manager has overall control over customer satisfaction and the profitability of the department. S/he normally supervises the Service Department staff, manages the sold and unsold hours, and meets with any dissatisfied customers. The objective of the service manager is to maintain high technician productivity, control expenses such as shop supplies and policy adjustments, increase gross profit, and promote high customer satisfaction scores and retention percentages.

Assistant Service Manager

The assistant parts manager reports to the service manager and assists the manager in the daily operations and supervision of the service employees. In a large dealership, this position might have responsibility for supervising the service advisors or acting as a shop foreman. The assistant manager replaces the manager during vacations and sick leave.

Service Advisor

The service advisor reports to the service manager or assistant service manager. This position is responsible for generating service sales during the write up process. The service advisor notes the customer's complaints and then checks the mileage against suggested maintenance schedules. If it is time for regular maintenance, then the service advisor attempts to sell that service. The service advisor is responsible to give the customer an estimate of the cost of the repairs and a time that the vehicle will be ready. It is the goal of the service advisor to have a high labor and parts sales per repair order and high customer satisfaction scores. Most service advisors are paid on a percentage of labor and parts sales.

Dispatcher

The dispatcher reports to the service manager or assistant service manager. This person is responsible for making sure that all service written is actually performed by the technician. It is the goal of the dispatcher to have the customer's vehicle ready when promised and provide high technician productivity and efficiency. The dispatcher maintains either a log of what work is being done by whom, or updates the computer with the information.

Shop Foreman

The shop foreman reports to the service manager or assistant service manager. The foreman is usually either a working foreman that receives his own flat rate hours and a small portion of the other technicians, or a non-working foreman that is normally paid a salary. The responsibilities include the training and supervision of the technicians. This position also includes maintaining the tool inventory and shop manuals, and scheduling classes. The shop foreman often does the quality control test drive of the vehicle before delivery to the customer.

Technician

The technician or mechanic is responsible for performing the repairs as requested on the work order. Sometimes only the problem is noted and the technician will have to perform diagnostic service to find what repair should be done. This may involve hooking it up to a diagnostic computer or test driving the vehicle. The technician is responsible for reporting back to the advisor if any additional repairs need to be done and waiting for authorization. After authorization, the technician requests the parts from the parts department and completes the repairs. Technicians are also required to maintain a tool inventory that they own and complete the required courses for certification.

Booker or Biller

The booker or biller reports to the service manager, assistant service manager, or controller. The booker or biller is responsible for adding labor and sublet to the repair orders. The booker or biller also flags the technician for the flat rate hours that they completed. This position is often combined with other jobs such as the warranty administrator, cashier, or dispatcher. This is the position that "closes" a repair order and prepares it for customer payment.

Cashier

The cashier reports to the service manager, assistant service manager, or controller. The cashier is responsible for customer payment, billing, and open repair orders. The cashier must provide a daily reconciliation of the repair orders and parts tickets that were paid for and the method of payment. The cashier balances the opening cash fund (drawer) to the ending drawer less payments received. The cashier is normally responsible for filing of repair repairs, collection of customer surveys, and processing of service reminders.

Warranty Administrator

The warranty administrator reports to the service manager, assistant service manager, or controller. It is the responsibility of the warranty administrator to apply for warranty reimbursement from the factory for repairs done on vehicles. The warranty administrator must follow the factory's warranty policy and procedure manual and attempt to claim the maximum amount of credit for the repairs. This includes careful attention to administrative and diagnostic time as well as the best price for parts and supplies. The warranty administrator maintains the warranty receivable schedule and must insure that claims are submitted and paid quickly. Any shortages must be investigated and reported to management.

Customer Relations Manager

This is a newer position to dealerships. During the past 5 years, the factory has required that the dealerships provide better customer satisfaction and retention. Many dealerships have added a QCP or CSI manager that acts as a liaison between the sales & service departments and the customer. The customer relations manager provides management with the results of a follow up call done usually 2-3 days after the customer picks up their car. The manager asks the customer if the work was done to their satisfaction and in the time frame expected. The customer relations manager attempts to satisfy any customers that are displeased with their service or sales experience.

Developing Long-Term Employees

One of the biggest personnel challenges that faces a service manager is how to keep good employees during the "slow" times. Most dealerships have high and low cycles throughout the year and from year-to-year. The typical service employee is highly trained by each manufacturer type and an experienced technician is often a premium. Productivity is measured by sales per employee and when a dealership experiences a downturn, the Service Department appears to be over staffed and nonproductive.

Productivity should not be confused with long hours, overtime and job related stress. These attributes may be present when a person assumes a new position but should, in this instance, be only of short term duration. The goal of all productive people is to find an easier and more effective method to accomplish certain tasks. The terms of each task should be reviewed and fully comprehended before changes are made. The process of change is an absolute, given our business environment.

When analyzing Service Department personnel for productivity, it may help to use some of the benchmarks below as a guide. One of the best solutions for expense control with regards to personnel management is to create pay plans that reward employees for productivity and customer service.

Quick Check

☑Run a service advisor's report to make sure that all the sales are being tracked by advisor. This report should contain both labor and parts and show totals by customer, warranty, and internal. Hopefully there is some portion of their pay that is based on this report.

Encouraging Employee Honesty

Internal controls (discussed in a later section) are an essential management function to protect against theft. Another method is to encourage employee honesty through example. It is difficult to expect service employees to be honest about the supplies when the dealer, other managers, or even the service manager take tools, manuals, and shop supplies out of the Service Department.

Providing Customer Service

Service Department employees should attend regular customer service training seminars. These employees spend more time on the phone and in person with customers than most other employees. It must be remembered that customers are not just the people who fill out surveys, but are also the other employees, the sales department, and factory personnel. It is important to establish a good relationship with the other departments. The service department depends on the sales department for future customers and it depends on the parts department for prompt ordering of their customer's parts.

Paperwork and Processes

It is important to study the process and flow of paperwork through the Service Department. After the addition of computers, many manual systems remained and were used parallel to the computerized system. In an effort to reduce expenses and increase productivity, each of the different flows must be analyzed. Steps that are redundant and do not contribute to either gross profit or internal control should be eliminated. Other processes need to be combined and performed by the same person for efficiency. Anytime a form can be eliminated it should. Since repair orders are very expensive to print, make sure each copy is being used. The other flows to study in the Service Department are:

- ☑ The flow of a parts request from the technician to the back parts counter, to the issuing of the part to the technician and entering the part on a open repair order. Make sure there isn't wasted technician time.

- ☑ The flow of an advisor's request for a part, a special order request being generated and notification to the advisor that the part is in. Make sure there is a system to track these orders and a method for the advisor to check on the status quickly.

☑ The flow of any sublet bills. Often the bill is stapled to the hard copy and is never forwarded to accounting or entered on the repair order. Review the sublet schedule to make sure all sublet is getting put on repair orders.

Quick Check

☑ Most open repair order listing have sublet repairs on them. Run this listing prior to your review of the sublet schedule. It will be easy to eliminate those debits that are for open repair orders. For the remaining debits, perform a history inquiry to see if the sublet repairs are a performed operation. If not, then pull the vendor invoice and purchase order and find the VIN#. Do an inquiry by VIN# to find the correct repair order number. Require that all vendors put a VIN# on their invoices.

Forecasting and Planning-Service Department

The following benchmark figures and comments have been obtained from studies undertaken on profitable dealerships. Some of the data is national in its origin rather than regional. It is important to forecast what the existing Service Department would have done based on current sales and with industry benchmarks. Some benchmarks can be used to create pay plans and offer incentives. Other are generalizations used to find if certain process problems exist or can be improved upon.

1. Hours per R.O. should be 2.8 - 3.5 - customer pay
2. Parts per labor dollar should be at least .90
3. Ratio of technicians to other service personnel - 4 - 1
4. Ratio of technicians to service advisors 5 - 1
5. 18 to 25 repair orders per day per service advisor
6. Service salaries should not exceed 35% of gross
7. Open R.O.s should not exceed 2 per Tech.
8. 10K to 12K parts and service gross per Tech.
9. Policy 2% of Gross
10. Labor Gross Per Service Advisor $30,000-$35,000
11. Total service personnel expense as % of service gross - 30-40%
12. Service Gross per Service Employee $3,500-$4,500
13. 25-30% net profit to gross
14. The percent of gross profit to sales is usually 70-75%.
15. The average hours sold per R.O. is 2.0 - 2.5

Effective Labor Rate

Although a dealer may have a stated labor rate of $55 an hour, they may offer specials and coupons that bring the actual (or effective) labor rate down. For example;

The dealer's labor rate is $60 an hour

1. 10 hours of customer pay work is done at $60 an hour
2. 10 hours of menu specials is done at $30 an hour
3. 10 hours of warranty work is done at $55 an hour
4. 10 hours of internal work is done at $40 an hour

The total hours are 40 and the total labor sales amount is $18,500. The effective labor rate was $46.30 an hour. There are numerous ways to obtain the effective labor rate:

1) Survey of repair orders. Collect a sample of repair orders and total the labor sales. Total the flag hours. Divide sales by hours to find ELR.

2) Computer listing. Obtain a listing of all labor hours flagged to technicians (Technician Detail Report). Divide total labor sales on the financial statement by these hours.

3) Payroll records. Find the total hours paid and charged to work in process from the payroll summary. Divide total labor sales on the financial statement by these hours.

Do not make a mistake and take the total technicians times the hours in the month to find labor hours. Technicians flag more hours than the shop is open due to 125% or more productivity. Be careful not to pick up sublet labor in total labor sales.

Breakeven Analysis for the Service Department

It is important that the service manager know what it takes to breakeven each month. The following is an example of a breakeven analysis:

Current Average Service Department Expenses	$89,423
Current Average Labor Per Repair Order	$148
Current Average Gross Profit Percent	75%
Current Average Number of Repair Orders	688
Labor Sales Per Technician	$7,250
Total Current Average Gross Profit	$ 76,124
XXXXXXXXXXXXXXXXXXXXXXXXXXXXXXXXXXXXXXXXXXXXXXXX	
Total Average Net Profit	$ (13,298)
Net Profit per Repair Order	($19.33)
Breakeven Point in Labor Sales	$119,231
Breakeven Point in Repair Orders	808
Breakeven Point in Technicians	16
Forecasted Repair Orders	1,000
Forecast Labor Sales	$147,528
Forecasted Gross Profit	$110,646
Forecasted Net Profit	$21,223

To find the breakeven point in labor sales, divide the total expenses by the gross profit percentage. To find the breakeven point in repair orders, divide the breakeven labor sales, by the average labor per repair order. The breakeven point per technician is the total breakeven labor sales divided by labor sales per technician. All amounts above the "XXXXXX" line are input and below are calculations. To find labor sales per technician:

Total hours worked by technicians (10)	1600	
Productivity %	125%	
Total labor hours		2000
Effective Labor Rate	$36 hour	
Total Labor Sales		$72,000
Sales per technician	$7,200	

Developing Pay Plans

The average compensation for a service manager is 5-10% of the service gross. Rather than paying strictly on the gross profit of the Service Department, some progressive dealers are paying on the whole dealership profit or the service and parts gross profit combined. Compensation plans should be structured to promote teamwork and inter-department support and cooperation. There are ways to structure pay plans that will make managers become more involved as team players.

A suggested method is to pay a service manager a percent of the combined dealership gross profit. A set percent of dealership gross profit would go into a pool. Each manager would have a share of this pool depending on their contribution, experience, and seniority. Another method would be to pay a service manager on both the service, body shop, and parts gross or net profit. This helps cooperation between the departments that rely on each other the most.

Paying Service Managers & Service Advisors

The important thing to remember when analyzing pay plans is to determine if the pay plan is within expense guidelines and the goals of the dealership. If it is the dealership's goal to provide good customer service, then this needs to be in the pay plan. On a more finite level, if it is the goal of the Service Department to increase hours per RO, then this needs to be rewarded in the service advisor's pay plan.

Some common service manager & advisor pay plans include:

- Commission on service & parts sales volume or gross profit
- Bonus for hours per RO
- Bonus for CSI or QCP scores
- Graduated commission based on total customer pay volume
- Percent of total gross profit less policy adjustments
- Bonus for high dollars per RO
- Net profit of the service & parts department

Work in Process

The most effective way to measure if the amount of work in process is reasonable is to convert the end of the month dollar amount into the number of flat-rate hours this balance represents. To do this you have to divide work in process by the effective hourly cost of labor. For example, assume the average technician's cost of labor is $14 per hour. If the dealership has $1,400 tied up in work in process, then this account represents 100 hours of work in process labor. The most accurate way to calculate the effective average cost of labor is as follows;

- Add up the total amount paid to technicians for the previous period
- Add up the total flat rate hours this represents.
- Divide line 1 by line 2 to get the effective hourly cost of labor.

Now it becomes easier to decide whether the amount of work in process is reasonable or excessive.

Divide the work in process hours by the number of technicians the dealership employs.

This will give you the average hours of work in process for each technician.

Taking the above example of 100 hours of work in process and having 20 technicians, the average hours per technician is 5 Service technicians should not have more than 2.5 hours of work in process outstanding each and body shop technicians about 10 hours. The following is a worksheet for determining excessive work in process.

Work in Process

Line#	Service		
45	WIP Service	$7,652.00	YTD
46	Average cost of labor per hr.	$18	Input
47	No. of hours in WIP	425	Line 45/46
48	No. of techs	10	Input
49	Avg hours per tech in WIP	42.5	Line 47/48
50	Budget hours per tech	2.5	Input (10 for body)
51	Excess in hours per tech	40.0	Line 49-50
52	Total (Excess) in WIP	($7,202)	(Line 51*48*46)

How to Control Work in Process – Get Rid of the "Lick and Sticks"

The most common way to control the work in process account is to schedule it. The debits on the schedule come from posting the daily time tickets, the reference number is the technician number and the date, the amount is the flat rate hours times the technician's cost, and the control number is the repair order number. The credits to the schedule come from the posting of repair order sales. The reference and control number is the repair order number and the amount is the cost of the labor sales (as a credit to work in process inventory.) If the account is not scheduled, then there should be a regular audit of the time tickets to the repair orders. This audit is called checking the technician's flags to the hard copies. After the ticket is complete, the cashier compares the daily time tickets to the front of the repair orders and back of the hard copies. All three amounts must be the same.

The best method of controlling work in process is to pay the technicians from a computer generated report instead of the daily time tickets. If the computer system is integrated to service invoicing, you run a report similar to the service advisor's report for each technician, and just pay them from the totals on that report. There is no longer a need to flag the technicians for their labor on those expensive daily time tickets. The benefits of using this system are:

- The amount that is charged to the customer or factory is the same amount that the technician is paid. No more charging the customer 2.3 hours and paying the technician 3.2 hours by mistake.
- The technician will always make sure that the work he has done gets entered on the computer. That is the only way he will get paid.
- One of the most expensive forms (the "lick & stick") is eliminated.
- The booker or biller will only have to flag the hours once - on the computer - instead of twice (once on the computer and again on the time ticket.)
- There is no clerical work required to convert the flat rate time on the time tickets to a dollar amount, total the time tickets, and post to the work in process schedule.
- There is no need to "check off" the flags to the repair orders for an internal audit (if this is even being done.)

To make this system work, the two methods have to run parallel until the technicians are comfortable with their computer generated report. After a few weeks of seeing that this report matches their daily time tickets, then you can suggest that they stop flagging. Technicians will still need to "punch on and off" of a job on the hard copy, since the factory requires to see actual time on jobs. The factory also requires that the amount paid to the technician be on the repair order. Instead of the flag on the back of the hard copy, the hours paid for the technician is on the front (since the dealership pays exactly what is computer flagged.) This is a modification of the existing ERO and ESI (Electronic Technician) software that is already approved by the factory for CDK and Reynolds + Reynolds customers.

Open Repair Orders

One of the most important reports to run in a dealership's service department is a listing of open repair orders. If the dealership has computerized service invoicing, then you need to request the following reports:

CDK	MR RWP
Reynolds + Reynolds	3613

It is also important to run a report of those repair orders that are closed in service and are waiting to be updated to accounting. They are:

CDK	DR RAC
Reynolds + Reynolds	3614

These lists will indicate which repair orders are still being worked on or have been completed but not paid for. Many on the list could be for vehicles that have been taken by the customers without payments. It is important that an explanation be provided for each item on the list.

If the dealership does not have computerized service invoicing, then you will need to create the list manually. Since the open repair orders are normally kept in two places; the dispatch office and cashier, this is a good place to start. You need to find all the hard copies and list any labor or sublet that has been done, the customer's name, and status. The following worksheet might help you create this listing. The dealership should create a listing of these repair orders at least monthly to balance the work in process account to. They may call it the work in process listing or open repair order report.

Reconciling Work in Process

The total of labor at cost from the open repair listing should balance to the amount in the general ledger. You will need to make sure there was a clean payroll cut off - the accrued technician pay must have been entered for any split pay periods and all repair orders must be updated to accounting. For example, if the dealership directly charges work in process from payroll, then make sure that either the pay period ended on the date of the open repair order listing or an accrual was made for any daily time tickets not paid for.

Investigating Shortages in Work in Process

The worst reason why work in process may be short (the balance in the general ledger is greater than the total of the open repair order listing) is that the dealership has a problem with technician labor theft. This is actually a very easy thing to do if the dealership has no internal audit control. The steps involved are:

1. The technician takes his daily time ticket and flags himself for 3.2 hours with repair order number 3604.
2. Either the technician flags the hours on repair order 3604 as less (if he actually worked on the job- let's say .2 instead of 3.2) or if this is a fake repair order, he destroys it.

To monitor this, it is important that someone "check off" the daily time tickets to the repair orders or use the process above to eliminate the "lick and sticks."

The next reason why there may be a shortage is simple human error. The booker or biller makes the same mistake as above, pays the technician 3.2 and only enters .2 on the repair order. Due to human nature, any error in the technician's favor is normally not reported to the booker. The same monitoring must be done as above. Another problem exists when the wrong costing is being done by the accounting office. If the technician is being paid $14 an hour, but the person who is doing the costing is using $13, then a shortage will accumulate. Make sure that the person responsible for costing (or the computer setups) has the correct payroll rate for the technician. The third reason is dealership theft of the cash sales. If an employee is keeping the amount that the customer paid and destroying the ticket, the work in process inventory will never get relieved. Both the above method and a missing document report will monitor this. The final way that work in process gets out of balance is improper account distribution by the payroll department. If a technician has a total of 174 hours and 8 of it is training time, and another 16 is sick leave, then only 150 hours should be charged to work in process. A careful review of the daily time tickets against the payroll journal could discover these errors. Often the training time is entered in the middle of a time ticket and not broken out separately at the bottom. Make sure that this is not happening.

Shop Supplies

The key to getting the best price for shop supplies is to get written bids from three different vendors. There should be a form letter made that invites the vendor to bid. The bids should have a price protection clause for a least one year. After the vendors are selected there should be a department vendor guide for each manager. Periodically, there should be an internal audit of the prices compared to the bids. Be careful on 52 week contracts for uniforms. Since technicians work only 45-50 weeks due to vacations and training time, this contract needs to be adjusted. Other shop supplies such as lubricants, nuts & bolts, and towels should be purchased by the parts department and issued on a ticket by technician number. At the end of the month, these supplies should be charged to the shop supply expense account. Most dealerships charge the customer for shop supplies on each repair order. It is important that the credits from these repair orders go into a sub account so the total debit expense is not distorted by the credits.

Shop Tools

Shop Tools are normally the tools that the dealership buys that are for a specific vehicle model. Although the technicians own most of their own tools, they cannot be expected to buy these special tools each year when a new model comes out. Unfortunately, these tools are expensive and difficult to control. It is important that only the service manager approve the purchase of these tools and that he doesn't automatically take every tool that a factory ships to him. Sometime the factory assumes that you might need a certain tool, but if they never work on medium sized trucks due to hoist restraints, it is unnecessary to keep the tools that are required for these models. Make sure that all shop tools are kept in a secure area and there is a check out log for the tools. To review shop tool purchases, look at the parts statement file for the factory. Most shop tools come on a parts invoice or other factory billing.

The Parts Department

The parts department is the most detailed and complicated of the gross profit generating departments of a dealership. It usually has one of the smallest gross profit contributions to the total dealership gross profit, but if managed well, the parts department can contribute an excellent net profit. The key is to keep the expenses low and manage the inventory to prevent shortages and obsolescence.

The dealership parts department is usually located in a secure area of the dealership and often attached to the service department. The area is secure due to the amount of physical inventory kept on hand and a history of theft in this department. It is attached to the service department because this is often its biggest customer. If you look at the dealership's financial statement, you can determine where they do the majority of their business.

- Parts counter sales are usually made to local mechanics, do-it-yourself owners, and employees. The gross profit percentage is between 35-40%.

- Wholesale sales are sales to other dealerships, part retailers, and body shops. The gross profit percentage is between 20-25%.

- Internal sales are created when the dealership adds accessories to vehicles like floor mats, reconditions used-vehicles or sells supplies to other departments. The gross profit percentage is between 15-20%.

- Repair order sales are sales of parts on repair orders in the service or body shop departments. The gross profit percentage is between 35-50%.

Counter, wholesale, and internal sales are usually generated on a parts invoice or ticket. Repair order sales are created on a repair order or work order. Most computer systems control these sales internally, meaning that the part transaction is stored in the computer system. The system generates the invoice or ticket number when the ticket is opened. The system relieves the inventory as soon as the quantity is entered, regardless of whether the sale is completed or not. The part remains as "work in process" on an open invoice or open repair order listing. To determine if the dealership uses this system, look at a repair order. If the parts are written on the repair order by hand (manual system, not computer generated) then there should be some indication next to the part that it has been "posted" or relieved from the inventory system. If they are still on a manual system, then during a physical inventory, these parts must be listed and totaled by hand.

There are two different computer programs controlling the parts department. The parts inventory system that controls the physical inventory, production of the inventory pad, stock orders, and parts receipts. The parts invoicing system, that was developed later, controls the sales of parts, invoicing, customer pricing methods, sales tax methods, and the accounting interface to the general ledger. The invoicing system usually automatically posts the sales transactions into accounting based on the setups in the accounting interface.

Quick Check

☑Make sure the parts department is a secure area. The back door should be locked from the outside and employees should not park near it.

☑Run the open invoice and repair order listing to make sure parts "work in process" is not excessive or overage.

DAILY OPERATIONS in the Parts Department

Inquires for Parts

Most parts requests come from the back service counter. A technician completes a part requisition form and the counterperson begins the process of finding the part number. Sometimes they have the part number if the job is a recall or regular maintenance item, but often the counterperson must rely on parts catalogs, microfiche, or an electronic parts catalog to locate the part. The procedure with the parts catalog and microfiche are basically the same; the counterperson find the book or film for that type of part and then looks at diagrams to visually find the part. With an electronic parts catalog the procedure is simplified; the year, make, model of the vehicle is entered and then almost any other piece of information to find the part. For example, to find a water pump; they can enter wat..., pum.., or cooling system.... The options that are available display and the counterperson can continue selecting until the desired part is found. At any time, the diagram can be printed or displayed for the technician. When the part is selected, and if the parts catalog is tied into the in house computer system, the availability is displayed. Some computer systems are also tied into a locator system that attempts to find the part locally if it is unavailable. If the part is unavailable and not able to be located, then either a lost sale is entered or the item is special ordered. Some computer systems include the special order function as part of the lost sale entry and others require that it be ***either*** a lost sale or special order. Make sure you follow the steps required by your computer company to avoid registering two sales for each special ordered part. If the part is available the counterperson enters a quantity, then a bin location is given and a picking ticket printed. This ticket enables the counterperson to make one trip to the shelves to collect all the parts. Once the quantity of the part is entered, most computer systems relieve the pad inventory of the part and it becomes "work in process". If it is going on a repair order, the list of open parts can be printed from the service system. If it will be sold on a parts invoice, then there is often a report of open invoices available.

Quick Tips:

☑Installing an electronic parts catalog could save time and costly mistakes of ordering the wrong parts.

☑Often the parts requisition form is three part and expensive. The back counter can usually use a one or two part form instead.

Selling Parts

To complete the sale of the part, the counterperson will need to enter the customer information, the counterperson number, and sometimes the price level or price matrix code. Some systems will allow the counterperson to override the sales price and even the cost, but these overrides will be listed on an override report. If the part requires a core charge (see section on cores), then that will display also. To finish up the invoice, there are options to add freight charges, shipping and purchase order information, and payment methods. The payment methods are set up in the accounting interface and also post directly to accounting if the dealership has the parts invoicing software. After completion, the invoice prints. Normally three copies are printed and the flow of these copies is diagramed later in the text. An invoice can usually be reprinted, but it will indicate that it is a reprint.

Quick Check

☑Be sure to run the counterperson override report and review any unusual changes. If the number of overrides is excessive, you might want to look at the pricing matrixes.

☑Beware if most of the cash accounting copies are reprints. If a dealership sells a lot of common parts, a counterperson could print the original and collect the cash, then print a reprint for the next customer and steal the cash.

Cash Sales

It is common that the parts department will sell a part to a customer that does not have a charge account and has not paid for the part yet. An example of this is a local gas station that calls and asks for the part to be held for them. The gas station does not have an account and the counterperson wants to have the invoice ready for the customer when they come in. The invoice is processed to the "cash sale" account and usually remains on the "will call" shelf with the part. Sometimes the invoice and part are sent out with the dealership's parts delivery driver. For control purposes, a copy of the invoice should always remain with accounting. Now that there are integrated parts invoicing computer systems, this invoice is often automatically downloaded into the accounting general ledger. This causes the cash sale account to have a debit balance. Trying to find which invoices are not paid is a difficult task, especially when this is compounded by the normal accounting errors of having a charge in the cash sale account and vice versa. As one solution, split the cash sales clearing account into one for service and one for parts. Set up the parts sales cash clearing account to be scheduled and controlled by the parts ticket number. Every day the cashier will send up to accounting a summary sheet of all the money collected for parts sales, by parts ticket number.

Take this summary sheet and post it into accounting, thus relieving all the tickets represented by cash and creating a receivable that will be washed out by the cash receipt made out for the cash received. Run this schedule every day and investigate items that are more than a few days old. I believe that having the cash sales clearing zero everyday is an ambitious and not altogether realistic approach, especially if the dealership is doing a big volume. Having outstanding balances in the daily cash sales clearing account is okay, but these balances better be small, they better be current and they better clear shortly thereafter.

Quick Check

☑In order to monitor this account more closely, I suggest that these cash sales clearing accounts appear on the DOC or MIS report as separate line items. Then, at any time, the dealer and/or general manager can ask for a printout of the balance.

Approving AR Accounts

Most accounts receivable accounts originate in the parts department. The customers expect to charge parts since they will be making many purchases each month and one statement simplifies payment. The parts department normally gives the potential charge customer a credit application which is completed and returned to the accounting office. The accounting office calls a few of the references on the credit application for payment history. They then assign a control number to the customer and add them to the customer database. A separate computer job is usually performed to enable the customer to charge and a credit limit is established. It is suggested that there be a policy in force with regards to required payment history, credit limits, and periodic reviews. Each computer system has a method to review sales and payment history. Careful attention should be paid to how many returns are made, gross profit percentage, and payment dates. If for example, they have established credit for a body shop that returns over 50% of their sales, has a gross profit percentage of less than 25% and is always 60 days late on payment, then this account should be reconsidered.

Lost Sales

The accurate tracking of lost sales ensures a more realistic tracking of the potential demand for specific part numbers and improves the dealership's ability to purchase parts in anticipation of demand. The quicker a part becomes a stocking part, the quicker the dealership will improve its gross profit through higher stock order purchases rather than pay a premium to buy these parts through emergency purchases. The difference in cost between purchasing a part by a stock order or through emergency purchases is substantial. If you take all the benefits of a stock order into account the difference between a stock order and an emergency purchase is more than 35%. (See section on stock orders.) Look at all lost sales every month in order to ensure that the parts department is tracking these items. This in turn will lead to better stock order performance and improved profitability. The procedure to enter a lost sale depends on the computer system. Normally, an inquiry line is displayed and when the item is not available, an "L" is entered on the line for a "lost sale". The computer system counts this as a sale for stock order purposes. Another lost sale is when the price is too high. It is suggest that counter people keep track of fast moving parts that customers say they can get for less at a local parts house. If they lower the margins on these parts only, they might gain a few more customers. They could add a new customer type to track the results of the new pricing matrix.

Quick Check

☑Lost Sales are listed on the parts management report. Run this report to see if lost sales are being recorded. If it is low, run a lost sale report by counterperson to find out which counterperson is not recording lost sales.

Special Order and Non-Stock Parts

When a customer needs a part that the parts department does not stock, this becomes a special order parts request. The request is generated either by the retail counterperson, wholesale counterperson, or the service department. The parts department employee either adds the request to the computerized special order parts system or writes up a 2 or 3 part request form. The hard copy of this form is usually a postcard that can be mailed to the customer when the part arrives. It is advisable that the dealership use the computerized special order parts system. The basic system works as follows:

The counter person does an inquiry for the part:

- If the part is not in stock, a function is entered to go to the SPO screen.
- On this screen, more information is given such as customer name, repair order number, phone, etc.
- The SPO record is downloaded to the stock order.
- When the part comes in the SPO is updated.
- The customer is invoiced for the part.

It is important to use this system for two reasons; first it ensures that the part is tracked by the computer system and that the part is treated like a lost sale and counts towards phase-in criteria for the next stock order and second, in order to have a way to receive a report of the special order parts on hand. A lot of parts departments don't like using the computerized SPO system because it is easier just to complete a postcard order form and manually track the parts, but eventually many manufacturers will eliminate the annual or quarterly returns. Instead they will only allow the dealer to return a part within 30-60 days. With this new system it will be important to know which special order parts are close to being 30-60 days old so they can be returned for credit.

A report of all unclaimed special order parts should be run daily for the service advisors. They need lots of time to call the customer and arrange an appointment. The parts department should also show the advisors how to do an inquiry on a special order part so they can help the customer know if their part has come in, without transferring the call to the parts department. This will help ease the phone load in the parts department and provide better customer service. Statically, approximately 80% of parts obsolescence comes from special order parts that are never picked up. Since the nature of these "special orders" will not generate enough demand to justify stocking the part, it remains in the inventory system as a "non-stock part". Another reason for having non-stock parts is through the receipting of "factory recommended" stocking lists. These lists are usually generated for new model vehicles based on demand for similar parts in the past. Typically, this type of "non-stock" part will achieve stocking status in a relatively short period of time.

Dealers should not have more than 5-8% of their total investment in non-stock parts. How do you find the percentage of non-stock parts in inventory? Run the "parts management report" from the parts inventory computer system, and look at the section titled "investment". This section will have a category for non-stock parts. If the percent for non-stock parts is over 10%, the following actions should be taken:

1. Request a list of non-stock parts and review the list with the parts manager to find the nature of these parts. If these came from "factory recommended" stocking lists, consider qualifying the list next year based on the dealership's parts history.

2. If the reason is an excess of customer special order parts, the policy of deposits, notification, and returns to the factory should be reviewed. Customers should provide a deposit equal to the cost of shipping and handling of the part. A follow-up call system should be implemented to notify the customer. Parts not picked up after 30 days should be returned to the factory.

The most important analytical tool is monitoring the parts department's "pick up" percentage. It is a very easy calculation. If the parts department correctly codes all special ordered parts purchased and all special ordered parts sold, the dealership has the ability to easily monitor all special ordered parts transactions. The pick up percentage, or what some refer to as "special order sales efficiency", is the monthly special order sales as a percentage of monthly special ordered purchases. I suggest that you calculate this percentage using two-month moving average in order to smooth out this number. What should the dealership's pick-up percentage be? It should be above 90 percent. Anything less than this percentage will create obsolescence return problems as the special ordered parts not being picked up, plus the natural amount of creeping obsolescence in the inventory, will exceed the earned return allowance that is automatically being accrued through stock-order purchases. (Nissan - and a limited amount of other manufacturers - have a program that allows all special ordered parts to be returned to the manufacturer within a certain period after receipt without affecting their return reserve allowance.) In order to accurately calculate this percentage ALL special order parts must be coded correctly. This means that special order parts that are ordered as part of a stock order should be coded as being special order parts.

PURCHASING PARTS

Outside Vendors

A parts department faces two objectives; the first is to keep the parts that customers need on the shelf. This is done through stock orders. The second objective is to fulfill the orders of parts not in stock for customers that cannot wait until the part is back in stock. Most of these are considered emergency purchases for customers whose vehicles are broken down in the service department. The parts department purchases these parts from various outside vendors; other dealerships, parts houses, and remanufactures. Other dealerships are usually used for non-make vehicles such as Ford parts at a Chevrolet dealership and same make parts. The other dealerships normally sell the parts at 20-25% over cost, which is considered a trade price. Parts houses such as NAPA often sell parts that are similar to factory parts but made by another vendor. These parts are usually sold at less than factory cost and a discount is taken by the dealership upon purchase. Remanufacturers sell used factory parts that have been restored to like new condition. Again, these are often sold at less than cost and a discount is taken. The reason behind this discount is that all the cost information for factory parts comes to the dealership on master tapes from the computer companies. To attempt to average the cost each time a part is purchased would be a big chore. If a dealership buys the part from another dealer at trade price, or 25% over cost, then the dealership must either take a negative discount on the purchase or adjust the cost manually on the repair order or parts invoice. If the part is being sold on a warranty claim, the cost cannot be adjusted. That would cause an error on the submission of the claim to the factory. Instead, the entry must be made on the purchase invoice.

Quick Check

☑If you have a shortage in the physical parts inventory, check to see if the cost was adjusted on parts purchased from other dealers at a markup. Pull the accounts payable folder for a dealer and back track the invoices to the repair orders. Verify the cost amount or look for an adjustment on the invoice.

As technology increases, the chore of acquiring a part gets easier. There are parts locators that query other dealers for the part that is required and some parts houses provide modem access to their inventories. Regardless of the methods used to locate the part, it is still a time consuming chore to acquire the part. The parts department must:

- Find the part
- Issue a purchase order for the part
- Send a runner to pick up the part
- Notify the service department that the part has arrived
- Receipt in the part in the inventory system
- Approve the invoice for accounting and attach the purchase order
- Sell the part on a repair order or parts invoice

Stock Orders

Stock orders are the regular order of parts (usually weekly) from one manufacturer based on selling trends, current days supply, and phase-in criteria. The first time the parts department buys a part, it is added to the inventory system as a non-stocking part. An example of a phase-in criteria is:

After **2** sales in **6** months - change the status from non-stocking to stocking.

When they run the next stock order report - if they have sold 2 parts within a 6 month time, **and** they are under the days supply guide - then it will put this part on the next stock order. Each computer company uses a different type of phase-in criteria. The key is to find the "source control" for each manufacturer and edit the criteria. Logically it is easier to fill the parts request from a part that is sitting on the shelf, rather than to send a runner to another dealer to get the part. Not only is there less labor and time involved, the price that they pay for the part will be less. It is the objective of a parts manager to order 75% or more of the parts through the stock order process. The process of a stock order includes:

- Request a suggested order report from the computer.
- Make adjustments to the suggested order based on the manager's knowledge. He might also add some special order parts or emergency orders to the order if the order is expected to be received in time.
- Authorize or confirm the stock order.
- Print the stock order.
- Mail or transmit the stock order via the computer.
- Receipt the stock order in the computer with exceptions upon arrival.
- Approve the invoice and forward to accounting.
- Stock the parts in their bins.

Stock order benefits may include:

Stock order discount from the vendor
Obsolescence allowance
Free freight
Warranty
Original equipment
Control
Easy return

The price of a part ordered on a stock order can be considerably less than purchasing it from another dealer or local parts house. The analysis on the next page illustrates the difference in the cost structure between stock orders and emergency purchases. (Source MNI)

STOCK ORDER			EMERGENCY
1.00	COST OF PART		1.00
.00	COST + 10%	.10	

(.05)	STOCK ORDER DISCOUNT	.00	
(.08)	OBSOLESCENCE ALLOWANCE	.00	
.00	FREIGHT (12%)		.12
.00	PHONE (long distance calls would increase cost)	.10+	
.00	LABOR ($4.00 hr/runner)	.20 (3 min)	
.87	TRUE COST		1.52

Measuring Stock Order Performance

Because there is such a large price advantage in buying parts via stock orders, it becomes important to monitor the purchasing performance. There are two independent methods that can be used to obtain this information and to verify its accuracy.

1. The parts management report.

The parts department prints out a parts management report every month. It generally contains a line that clearly indicates the stock order performance. This figure is generated as a direct result of how the parts manager coded all the parts receipts (purchases received by the parts department) for the month.

2. The accounting records.

These records keep track of all parts purchases made. This information, coupled with the parts summary report received from the manufacturer, will be sufficient to give you the information needed to calculate the stock order performance. Which of the two is the more reliable source of information? In my opinion, it is the accounting records and here's why:

The parts management report is open to possible errors, intentional or accidental. This affects the stock order performance calculation. For example, if a part is incorrectly coded as a stock order purchase when it actually should have been coded as an emergency purchase, then the stock order performance will be overstated. Therefore, relying solely on this report may be inadvisable.

How can you check the stock order performance via the accounting records? It is simple and here is how you do it. There are two steps to follow:

1. Isolate the stock order purchases for the month. Every manufacturer sends to each dealership a monthly parts purchases summary report. This summary report will split up all purchases made, one of which is stock order purchases.

2. Isolate the total purchases for the month. To do this, simply look at the summary of all purchases made to the parts inventory account in the general ledger for the month. The general ledger activity will be categorized by various sources. The sources you want to look at are "general purchases" and possibly "cash disbursements."

Let me illustrate how easy it is to use the accounting method to calculate the stock order performance. Let's say the parts summary report, received from the manufacturer for the month, indicates that they purchased $100,000 worth of parts from the manufacturer, of which $75,000 was through stock order purchases. By looking at the parts inventory account in the general ledger, you see that the dealership purchased a total of $150,000 worth of parts for the month. Now, by merely dividing the $75,000 into the $150,000 you find that the stock order performance is 50 percent -- clearly a percentage that needs to be improved. This percentage should agree with the parts management report. If it doesn't, then go by what the accounting records show instead. They are usually more accurate than the parts management report. Also, try and find out why these two reports do not agree. It normally is in the coding of the parts receipted into the parts system.

A suggestion: There may be a timing difference between when the parts department receives its orders from the manufacturer and when the accounting department processes these invoices. To smooth this out, I suggest that you use the past three months' purchases and stock order summaries. This will be a little more reliable.

Another way to track the stock order purchases:

Here's another way to isolate all the stock order purchases. Some dealers open up a "special inventory" account in the general ledger that will be used only for stock order purchases. The allocation in accounting is easy to facilitate and at the month end it becomes an easy task to capture this information from the general ledger. You can have this sub-account "point" to the MIS or DOC.

At the beginning of the following month, the closing balance of this "stock order inventory" account is then consolidated with the main parts inventory.

Understanding Best Reorder Point or Low Days Supply

If a dealership's inventory keeps growing without a significant increase in sales, there is a chance that the Best Reorder Point and Best Stocking Level are set too high. This term varies by computer system, but it is also considered the "Low Days Supply" and the "High Days Supply". Often a dealership will have 45 days as the low and 60 as the high. What this means is when the part quantity falls below a 45 day supply, it will order enough parts to have a 60 day supply on hand. The question to ask the parts manager is "How many days does it take to replace a part in this source?" Rarely will the answer be 45 days. Most parts are restocked within 3-7 days after being ordered. To maintain a 45-60 day supply of parts is often excessive. The best way to find the best reorder point is to slowly reduce the days by 5 each week until you reach a logical level. What is the logical level? I would suggest taking the maximum number of days between stock order and adjust it depending on how well the manufacturer fills that order. If the factory consistently fills 95-100% of their orders, then increase the days by only 50% as a buffer. A 7 day stock order would have a best reorder point at 11 days. You could double that just to make sure and keep a 22 day supply. You would slowly reduce the 45 best reorder point down to 22 days and watch the level of service. If the parts department is still filling the same percentage of orders, then the new setups are working.

Best Stocking Levels or High Days Supply

Most manufacturers do not provide a volume discount for ordering large quantities of parts. Based on this fact, there is no benefit in having a great variance between the best reorder point and the best stocking level. The only advantage would be a reduction in labor; it is sometimes easier to stock 2 parts on one day than stocking 1 part this week and another next week. If this is the case, then maybe a 25-50% spread between the best reorder point and best stocking level is suggested; a 22 day best reorder point would have a stocking level of 28-33 days.

This is the chart to calculate best reorder point and stocking level;

Number of days between stock orders
\+ Number of days to review and issue stock order
\+ Average number of days to fill the order
\+ Number of days to put away the inventory =

____
Total lead time
X 2 =

____
BEST REORDER POINT
X 1.25 =

____
BEST STOCK LEVEL

Understanding Phase-in and Phase-out Criteria

The phase-in and phase-out criteria determine if a part will be a stocking part or not. If a part is a stocking part, then it will be included on a stock order. To find out what the phase-in and phase-out criteria are on an CDK system, run job AM MNT. On a

Reynolds + Reynolds system, the job is 2323. As a guideline, a parts department should be as responsive to change as it can afford. Normal setups are 3 demands in 12 months to phase-in. With careful management, this can be changed to 2 demands in 9 months. That means that after the 2nd request, it will change to a stocking part. The phase-out criteria should also be as responsive as it can be; no demand in 6-9 months would be more aggressive, but would need careful management attention to avoid phasing out seasonal parts.

Economic Ordering and Cost Breaks

Most of the major computer systems allow the dealership to carry extra days supply of parts based on cost. If the cost of a part is small, let's say less than $1.00, then a dealer can usually afford to carry a high days supply of these parts at little additional cost. The dealer can also select a negative amount of days supply for higher priced parts. The setup for these cost breaks are in the same function as phase-in and days supply. The CDK term for the table is "Cost Breaks" under function 2 of MNT. Reynolds + Reynolds has an "Economic Ordering Parameter" that prompts after the phase-out criteria in job 2323. Power R+R (UCS) let's the dealer select on the PIC Vendor Option job, under MISC, small buy criteria. As a suggestion, any part under .50 should have an extra 30 day supply. This eliminates the nuisance of stocking these small parts. To determine the other cost breaks, run the piece ranking reports mentioned later and discuss with management. To recommend a negative days supply for a higher cost part is difficult. This is a risky determination, since the day supply should be accurate for all parts, especially high cost parts.

Understanding Stock Order Sources

The above items: days supply, phase-in criteria, and economic ordering breaks are determined by the setup for a particular source for purchasing the parts. Originally most dealerships had just two sources, the factory and other. A GM dealership would have source 100 for GM parts and 101 for other parts. As dealerships and franchises grew, parts departments started tracking many sources. Some creative managers split up the sources for each manufacturer into various sources. Since the days supply and phase-in criteria can be determined by source, fast moving GM parts could be in source 100, target GM parts in source 101, electrical GM in 102, body shop GM parts in 103, and so on. If a manager truly needs the customization of each source, then this method is advisable. Unfortunately, the parts management reports summarize each source individually and then as a total. If you want to see the performance of all GM parts and the parts manager has 10 sources for GM, then each would have to be manually subtotaled. If you find that all 10 sources have the same setup and no other reason for being split up, you may suggest combining them back into one source. To find out what sources a dealer has and all the setups, print a Specifications Listing, job 2329 for Reynolds + Reynolds. On an CDK system, display in job MNT.

Purchase Orders

A purchase order is an authorization to buy the necessary part, supply, or sublet repair. The forms are normally kept in a spiral bound book, three on a page, with three copies. The normal sequence of the copies are:

Original (white) Copy for vendor
2nd copy (yellow)Attach to vendor invoice for accounting
3rd copy (pink) Kept in book for reference

The purchase order contains; vendor name and address, quantity of items purchased, description, price each, extended price, accounting codes, and authorizing signature. The person writing the purchase order will often include the repair order or invoice number that these parts are being purchased for. Some computer systems now have electronic purchase orders. These systems provide immediate memo posting for the purchase order, rather than waiting for accounting. For example, on the CDK system, if a purchase order has been written to add wheels and tires to a used vehicle, the estimated amount and description will show in the used vehicle record until the purchase order is closed and the invoice paid.

Direct Checks and Accounts Payable

After receiving the invoice and attached copy of the purchase order from the parts department, the accounts payable clerk determines if the dealership has a control number set up for this vendor or not. If this is an occasional purchase, or only one invoice for the month, the clerk will often issue a direct check before the month closes. The clerk will match the statement to the invoice and directly charge parts inventory and credit cash with the check. If this vendor has many invoices, the clerk will post each invoice as a debit to parts inventory and credit accounts payable for that vendor. The next month when the statement comes from the vendor, the clerk will match the invoices and accounts payable schedule to the statement and pay the vendor. If the dealership has batch accounts payable software, the clerk will enter the vendor control number and the invoices eligible for payment will display. The clerk will either validate the invoices or put a code to hold invoices that have not been billed yet. When the clerk prints the check for the vendor, the detail in the schedule clears.

Manufacturers Statement and Parts Summary

The manufacturers statement and parts summary is merely a very complicated vendor statement. Most of the transactions between the factory and dealership are detailed on this statement including floor plan credits, gas credits, warranty claim credits, rebate payments, charges for sign leasing, brochures, yellow page advertising, computer charges, and part purchases. To aid in segregating the parts purchases, the factory normally provides a parts summary. This summary lists each parts invoice for the month, the net amount, the amounts eligible for returns, and any freight and discount credits. The summary page normally discloses the year-to-date reserves for returns and any advertising fund balances.

Sublet Sales

A sublet sale is when the repair work or addition of accessories for a vehicle will be done by an outside repair shop or vendor. An example of this is wheel balancing. Often the dealership will not have the equipment to perform this type of work and they will send the work to a local shop. The service advisor sells the repair work and often calls the local shop to make an appointment. The work is performed, then the local shop takes the invoice to the parts department to get a purchase order. Unfortunately, the control over sublet repairs is handled in many dealerships by the parts department; and they don't get paid on sublet repair profit. The reasoning behind the parts department handling sublet is that they have the purchase order book and can authorize payment for sublet repairs. Unfortunately, they usually don't know what really needed to be done to the vehicle or when it needed to be completed. They often don't "mark up" the invoice when they transfer it to the repair order and the profit is lost. If the dealership wants to make money on customer pay sublet, they'll need to transfer control from the parts department to the service advisors and pay them on it. This means either giving them a purchase order book or giving the task to the service booker. If they have electronic purchase orders, the process is simplified. If they still want to leave the control in the parts department, they should consider paying them on customer and internal sublet repairs. Warranty sublet repairs are normally required by the factory to be sold at cost.

Rather than customize their pricing matrixes, many dealers have cost = sale for the default for all sublet repairs and are only sometimes manually marking up customer and internal sublet repairs. They can get their computer system to mark up sublet repairs automatically. For Reynolds & Reynolds users, the sublet markup is under the Make Billing Guide (3624). For CDK users, sublet is controlled by job OP UOC. There is a percentage markup field for each labor type. I would suggest that they start out with customer sublet at 40%, internal sublet at 25%.

Should they mark up internal sublet repair? The best way to get the new and used departments treated like customers is to charge the same for internal sublet repair as they do customer sublet repair and pay the service advisors on internal sublet repair. If they make a sublet repair profit of $100.00 in the shop, they'll normally pay an advisor and manager a commission of between $7.00 and $9.00. If that same profit goes back to the sales department, the commission is usually between $25.00 and $40.00. If the sales managers refuse this mark up, they might want to try to get a better price from their suppliers for internal sublet.

To find out if the dealership is marking up sublet repairs, look at the gross profit for sublet repairs on the financial statement. If there is little or none there, you might want to investigate further. Unfortunately, most financial statements group warranty, customer, and internal sublet together. They'll need to set up some sub accounts for each type of sublet sale and cost of sale; customer sublet repair, warranty sublet repair, and internal sublet repair. Then have them put these new accounts on the DOC to get the gross profit percentages by type. Now they can monitor if they are making money on sublet repairs.

MANAGEMENT in the Parts Department

Employee Positions and Job Descriptions

The normal positions in the parts department include:

Parts Manager
Assistant Parts Manager
Counterperson
Outside Salesperson
Parts Runner or Driver
Purchasing Agent or Inventory Clerk
Stocker or Receiving Clerk

Parts Manager

The parts manager reports to the general manager or the parts and service director. This manager has overall control over the physical parts inventory and the profitability of the department. S/he normally supervises the parts department staff, determines inventory ordering levels, and customer pricing procedures. The objective of the parts manager is to provide the internal controls to prevent shortages and obsolescence, generate gross profit, and control expenses.

Assistant Parts Manager

The assistant parts manager reports to the parts manager and assists the manager in the daily operations and supervision of the parts employees. In a large dealership, this position might have responsibility for the computer system, physical inventory, or wholesale operations. The assistant manager replaces the manager during vacations and sick leave.

Counterperson

The counterperson reports to the parts manager or assistant parts manager. This position is responsible for generating parts sales either through inquiry of the bin inventory or locating the part at another dealership or part vendor. The counterperson processes the sale from inquiry through invoicing and often receiving payment. Each counterperson's sales and gross profit amounts are monitored for productivity.

Outside Salesperson

The outside salesperson reports to the parts manager or assistant parts manager. This person normally handles the large wholesale accounts and body shops. Normally the outside salesperson visits the customers to check stock levels and procure a purchase order for parts. Sometimes the outside salesperson will deliver the parts to the customer and complete the invoicing procedures.

Parts Runner or Driver

The parts runner or driver reports to the parts manager or assistant parts manager. This person has two functions; the first to go get the parts that the parts department has purchased and needs immediately and the second to deliver parts to customers such as body shops, local gas stations, other dealerships, and part houses. The runner takes these parts with two copies of the invoices to the customer for either signature or payment.

Purchasing Agent or Inventory Clerk

The purchasing agent or inventory clerk reports to the parts manager or assistant parts manager. In a larger dealership, this position is responsible for processing the daily orders to the factory and weekly stock orders. The clerk might also be responsible for locating parts on an emergency basis, making the decision on whether to wait for a stock order, acquire the part locally, or request a VIP order.

Stocker or Receiving Clerk

The stocker or receiving clerk reports to the parts manager, assistant parts manager, or inventory clerk. This position is responsible for the receipt of parts shipments and placing the parts in their proper bin location. The clerk looks for shortages, damage, and incorrect orders. The clerk will often be responsible for shipping sold parts to customers, and processing the special orders.

Developing Long-Term Employees

One of the biggest personnel challenges that faces a parts manager is how to keep good employees during the "slow" times. Most dealerships have high and low cycles throughout the year and from year-to-year. The typical parts employee is highly trained by each manufacturer type and an experienced parts employee is often a premium. Productivity is measured by sales per employee and when a dealership experiences a downturn, the parts department appears to be over staffed and nonproductive.

Productivity should not be confused with long hours, overtime and job related stress. These attributes may be present when a person assumes a new position but should, in this instance, be only of short term duration. The goal of all productive people is to find an easier and more effective method to accomplish certain tasks. The terms of each task should be reviewed and fully comprehended before changes are made. The process of change is an absolute, given our business environment.

When analyzing a parts department personnel for productivity, it may help to use some of the benchmarks below as a guide. One of the best solutions for expense control with regards to personnel management is to create pay plans that reward counter people for working harder during the busy times. Another solution is the cross training of positions so a receiving clerk can help deliver parts and the parts runner can stock parts depending on the flow of work.

Quick Check

☑Run a counterperson report from the computer system. Make sure each counterperson is using their own number. Verify that this report is close to the total sales on the financial statement. If not, look for the GM internal parts transfer entry or parts being sold without going through the invoicing system.

Encouraging Employee Honesty

Internal controls are an essential management function to protect against theft. Another method is to encourage employee honesty through example. It is difficult to expect parts employees to be honest about the inventory when the dealer, other managers, or even the parts manager take jackets, hats, and T-shirts out of the parts department. It is important that the counterperson invoice these managers for the items and require an accounting.

Providing Customer Service

Parts department employees are normally left out when the dealership commits to a customer service training seminar. These employees spend more time on the phone and in person with customers than most other employees. It must be remembered that customers are not just the people who fill out surveys, but are also other employees, wholesale body shops, and factory personnel. If a counterperson is not helpful to an advisor who is checking on the status of a special order part, then the customer service for that person will suffer.

Quick Check

☑To find out if there is a customer service problem in the parts department, try a survey completed by technicians, service advisors, and other managers.

Reducing Expenses in the Parts Department

The largest expense in the parts department is personnel. This section will discuss how to reduce this with improved productivity and pay plans that compensate based on sales or gross profit. The following are some suggestions to reduce other expenses in the parts department.

Freight

The easiest way to reduce freight is to order most of the parts on a stock order. Increasing the level of service and true turns will provided this added benefit of lower freight charges. Make sure that all outgoing freight is charged to the customer and going back to the proper account number. Set up a sub account number in accounting for freight credits to monitor this procedure. Make sure the shipping clerk and counter people know the best way to ship items depending on the destination and weight. Sometimes using motor freight can save 50% over UPS on heavy objects.

Other Supplies - Parts

The bottom line on supplying low cost items such as nuts and bolts is that rarely does the specific cost of these items get posted to the repair order. The dealership normally absorbs this cost which, if not expensed monthly, is picked up and accounted for during physical inventory. To accurately keep tabs on small items leaving the parts department requires a lot of attention to detail and constant monitoring. One process is to create a specific range of part numbers that have a part description that appears on the repair order as "nuts and bolts". This range of part numbers should be limited and the number will be determined by the cost of the part. For example, you will have a specific part number that will control parts having a price under one dollar. Another number would control parts that cost between $1.00 and $2.99. (The part numbers could reflect the price range.) This system requires less vigilance and becomes an automatic process when placing parts onto a repair order. This will enhance your repair order yield as well as reduce the sloppiness inherent with these assets.

Many dealerships recover a certain amount of the cost of supplies on each repair order. This credit or recovery distorts or hides the actual cost of the supplies. If you are using this system I suggest that you create two accounts, one to track the cost and the other to track the recovery. These two accounts, although consolidated on the Operating Report, can be shown separately on the DOC. It will give you all of your expenses without the credits going to this account.

Quick Check

☑Make sure the nuts and bolt bins are not located in the shop for easy access. These should be behind the parts counter and controlled by the parts department.

Paperflow in the Parts Department

It is important to study the process and flow of paperwork and parts through the parts department. After the addition of computers, many manual systems remained and were used parallel to the computerized system. In an effort to reduce expenses and increase productivity, each of the different flows must be analyzed. Steps that are redundant and do not contribute to either gross profit or internal control should be eliminated. Other processes need to be combined and performed by the same person for efficiency. Anytime a form can be eliminated it should. The two paperwork flows outlined are common to normal dealership operations and can rarely be improved upon without further computerization. The other flows to study in the parts department are:

- The flow of a parts request from the back (service department) parts counter to the issuing of the part to the technician and entering the part on a open repair order.
- The flow of a phone/front counter request for a part, the issuing of the part to the customer, and invoicing of the part.
- The special order parts system (discussed in another section).
- The return of parts to the factory or vendor when defective or authorized by an allowance.
- The flow of cores, core credits, and returns (discussed in another section).

Quick Check

☑Make sure an accounting entry has been made for any returns that have been shipped. Most dealerships wait for the credit to be issued and received, but a safer and more accurate method is to make the entry as soon as the parts leave the dealership.

Forecasting and Planning Benchmarks – Parts Department

The following benchmark figures and comments have been obtained from studies undertaken on profitable dealerships. Some of the data is national in its origin rather than regional. It is important to forecast what the existing parts department would have done based on current sales and with industry benchmarks. Some benchmarks can be used to create pay plans and offer incentives. Other are generalizations used to find if certain process problems exist or can be improved upon.

- Sales per parts employee - at least $20,000 per month
- Total parts personnel expense as % of parts gross - 20 Group - 25%; NADA - 45%.
- 30% net profit to gross retainage.
- Gross profit to sales 35%.
- 60% of all parts sales will be to the service and body shop departments.
- Parts inventory equals 35% of the net working capital of the dealership.
- The holding cost of keeping stock on the shelf is 30%.
- A part in inventory for more than nine months without demand will seldom sell. If it does, the dealership will normally lose money on the transaction if all the incremental costs associated with this part are taken into account.
- More than 50% of the specially ordered parts will not be used or picked up.
- Over 80% of obsolescence comes from special ordered parts not picked up.
- There is a natural "creeping" obsolescence of between 7-10% annually.
- In the majority of cases, the additional money spent to obtain a part for a vehicle will never be recovered by the end user.
- Parts manager compensation as % of parts gross - 8%
- Experienced backcounter person per technicians: 1 to 10. After 10, a less experienced assistant for every 8 additional technicians.

Wholesale Part Sales

There is always a lot of controversy over the wholesale part business. To determine if the dealer should begin, expand, or eliminate a wholesale part operation requires a two part analysis. The first is to financially track wholesale sales - not by the standard account numbers since many retail sales go to a wholesale account number and numerous wholesale sales go to a retail account. The easiest way is to track these customers is with a special customer code and isolate the sales and gross profit by code. It is important to analyze the amount of returns done by these customers and their gross profit percentage. Most computer systems provide these reports by customer codes. The second analysis is the expense attributable to this department. A careful tracking must be made of personnel, shipping, and supplies. Other indirect costs should be considered such as the borrowing costs for a higher inventory value and obsolescence caused by high returns. Many of the incentives such as wholesale compensation are being eliminated as manufacturers change the way parts are priced. With this information the dealer can make a financial decision based on the profitability of this department rather than the prestige that a large parts volume brings to the dealership.

Developing Pay Plans

The average compensation for a parts manager is 8% of the parts gross. Rather than paying strictly on the gross profit of the parts department, some progressive dealers are paying on the whole dealership profit or the service and parts gross profit combined. This is considered a cohesive payplan. This helps encourage a shared vision of making sure customers feel comfortable and satisfied with the dealership and will come back again. To achieve this shared vision, it is necessary that all department managers have common goals and objectives. Many parts managers are out solely for themselves, their own employees, and their own parts department, rather than looking at what will be best for the customer and the dealership. Business is viewed from a vertical rather than lateral perspective. Finger pointing and turf-protection behavior appears more to be the rule of the day rather than formulating and then implementing a collective organizational strategy of working together as a team. Unfortunately these turf-wars hurt business, and the ability of the dealership to grow to greater heights, higher customer retention and enduring profitability.

At the dealership level, much of this vertical thinking mentality is self-inflicted. Unlike public companies, stock options are not available to a parts manager as a form of compensation, so it is difficult for this manager to look at the whole dealership from an equity position. The compensation method through only parts department profits cements the rather inward thinking decision making and creates inter-department rivalry. Ironically, it is the pay plan design that entrenches this narrow thinking focus. Compensation plans should be structured to promote teamwork and inter-department support and cooperation. There are ways to structure pay plans that will make managers become more involved as team players.

A suggested method is to pay a parts manager a percent of the combined dealership gross profit. A set percent of dealership gross profit would go into a pool. Each manager would have a share of this pool depending on their contribution, experience, and seniority. Another method would be to pay a parts manager on both the service, body shop, and parts gross or net profit. This helps cooperation between the departments that rely on each other the most.

The most common pay plan for a parts counterperson is a base plus a bonus or commission. The bonus or commission is normally based on a percentage of the following accounts:

- Gross Profit - Customer Counter Retail
- Gross Profit - Wholesale
- Gross Profit - Service Department

This traditional pay plan is okay if the parts counterperson is involved with retail and wholesale customers as well as dealing with service technicians. But if the parts counterperson works solely with the service department, then you may want to suggest paying the parts counterperson ONLY off a percentage of the gross profit derived from parts sales on repair orders. A variation to this would be to pay the counterperson on a combination of labor and parts gross generated in the service department. This kind of pay plan encourages improved focus on shop productivity and directly rewards the parts counterperson for achieving this goal. It also should lead to improved teamwork between parts and service employees. The important thing to remember when analyzing pay plans is to determine if the pay plan is within expense guidelines and the goals of the dealership. If it is the dealership's goal to provide good customer service, then this needs to be in the pay plan. On a more finite level, if it is the goal of the parts department to increase counter retail sales, then this needs to be rewarded in the counterperson pay plan.

Physical Inventory

It is important that the dealership take an annual physical inventory. There are numerous benefits to having an outside service perform this inventory including the internal control it provides and the reduction of overtime expense for the parts department. It often takes weeks to prepare for the physical inventory. All the part numbers must be checked to ensure that they have proper bin locations. All the bin locations must be properly labeled and accessible. The accounting office must prepare for a good accounting cutoff of purchases and sales. Normally the parts department will stamp all invoices "BEFORE INVENTORY" and "AFTER INVENTORY" to assist the accounting clerks. It is important that a worksheet be completed after the inventory to reconcile the physical count to the accounting GL. The form on the following page is an example of the worksheet:

PHYSICAL INVENTORY RECONCILIATION WORKSHEET

Physical Inventory:

Amount per physical count pad (attach pad)		$__________
Plus any parts counted but not on pad (attach a listing)		+__________
Plus any cores counted but not on pad (attach a listing)		+__________
Less any parts counted on pad but scrapped		-__________
Total Physical Inventory	(A)	$__________

Accounting Balance

Balance in Parts GL account acct#	___	$__________
Balance in any parts sub accounts	___	+__________
Other		___ +__________
Plus any purchase invoices not in acct. (attach a listing)		+__________
Less any return credits not received (attach a listing)		-__________
Less any work in process (attach a listing)		-__________
Less any invoices in acct. for parts not received (attach a listing and explanation)		-__________
Total Adjusted Accounting Balance	(B)	$__________
Adjustment necessary	**(A) - (B)**	**$**

Periodic Bin Checks

It is important that a dealership make periodic bin checks. A perpetual inventory system does not mean that an annual system is not necessary, but rather alerts a manager to any shortages or discrepancies before it becomes too serious of a problem. Most computer systems provide a listing of the inventory for a range of bin locations and it should become a daily routine for all parts employees to check a small range. Not only does this encourage careful restocking of the parts, but also makes employees aware that shortages will be routinely noticed. All adjustments should be made by the parts manager and investigated.

Finding Shortages

Often a shortage in the parts department comes from incorrect accounting. Here is a checklist to find shortages:

1. Check the parts reconciliation form. Are the items subtracted or added correctly (it's common to enter them in backwards.)
2. Look at the gross profit percentages on the financial statement – are they higher than other dealers of the same make?
3. When a dealer purchases a part from the outside that is for their make of vehicle, they must make a cost adjustment on the invoice to debit parts inventory for the tape amount and debit cost of sales for the markup from the other dealer. Is this being done?
4. Are there deletions being made on the parts pad? This normally indicates theft of the part.
5. Look at the override report – are they zeroing the cost on tickets? This normally indicates theft of the part.
6. Has appreciation been taken? See the section on appreciation.
7. In the accounting records investigate all checks and journal vouchers to parts inventory and any items from non-traditional sources (a debit to parts inventory and credit to cash sales.) This normally indicates theft in the accounting office.

Reports that Measure Inventory

Most of the major computer companies have a monthly parts management report that provides most of the information that you need to analyze the parts inventory. The following are a few guides that you may wish to use when comparing the information with that of a well managed inventory:

- Gross Inventory turns of above 5 times per annum
- True Inventory turn of 3.5 times per annum
- Stock order efficiency - or order performance - above 75%
- Level of service - or efficiency - of between 82% and 92%.

Run the parts management report from the computer system each month and pay attention to these lines:

- Inventory Balance - this is shown by number of parts and type of parts; stock order parts, non-stock, special order, etc.

- Inventory Aging - is shown by investment or movement. There are two ways to age the inventory; the first is based on how long since the last sale and the second is based on how long since the part was received.

- Sales, Returns, Receipts, Lost Sales, Adjustments, Parts with no cost, and Misc. items. These are good analysis tools for detecting problem areas.

- Ratios: Turns, true turns, level of service. These ratios are on the trend report and can be compared to last year or year-to-date averages.

ABC Report vs. Piece Ranking Reports

Most DMS systems provide many inventory ranking reports, the two most important are reports that measure the sales based on the cost of the part (ABC Sales Report or Sales at Cost Report) and the number of pieces sold (ABC Piece Ranking Report or Piece Sales Report.) Although the ABC Report has been a valuable measuring tool, sometimes the piece sale report needs to be included to get an overall picture. For example, if a dealership sells 10 engines a year at $1,500 each, the ABC Report will probably include this part. But if they sell 1500 filters at $3.00 each, these parts would probably be excluded. These parts would be on the piece sales report. Be sure to run both reports, for CDK these reports are in the English Report Generator: SRR and ABC. The Reynolds + Reynolds jobs are 2233 and 2234.

Parts Returns

The manufacturers often allow a certain amount of parts to be returned for credit. This return allowance is usually based on a percentage of stock order purchases and is indicated on the monthly parts summary report. Other manufacturers have a set return period and they send a notification to the dealer of the amounts and dates of the allowed returns. The parts manager usually requests a report from the computer system of parts that are eligible for return and edits this report based on the allowance given by the factory.

The accounting department should keep track of all parts returns made to the manufacturers at the time the parts are actually returned rather than waiting for the parts credit to be received. There are a number of reasons why the parts return should be accrued:

- The inventory has physically left the dealership and the financial records should reflect this.

- The parts management report - the most important parts management tool the dealership has - reflects the physical return of the inventory. The parts management report number for dollar in inventory will keep "in step" with the inventory reflected in the financial statement.

- This accrual will lead to a more accurate analysis of the dealership's calculation of inventory turn or days supply.

- Setting up a receivable for this return will allow the dealership to keep track of the amount owed. This is important because if the dealership does not receive full payment, you can ascertain reasons for the shortfall.

Quick Check

☑Are the parts that do not receive credit going to come back to the dealership's inventory or are they going to be scrapped? (If the parts come back into inventory, they should be re-entered using their original parts history data. If the parts are scrapped, then the cost of these parts should be charged against the parts department gross.)

When the parts are returned to the manufacturer, the accounting entry is to credit parts inventory and to debit a factory receivable. Please make sure to isolate the freight or other charges that may have been deducted from the credit received from the manufacturer.

Price Update Tapes

Usually every month, the computer vendors send a manufacturer's price update tape to the dealer for each franchise that they carry. This tape is loaded onto the computer then a price match is done to update the inventory cost and list price with the amount determined by the factory. This update usually results in an increase in the inventory value which is termed "parts appreciation". A report is generated after applying the update tape that provides this appreciation amount. It is important that these tapes be loaded upon receipt. Some computer systems require that the system be "down" when the update is done which makes it difficult to get the job done. Make sure that there are not any piles of unloaded tapes waiting to be updated.

Making an Accounting Entry for Price Increases

The price increase or update received from the manufacturer will not be picked up in accounting unless the accounting department specifically accrues or responds to the entry. The amount of the price increase can be picked up from the computer printout that is generated immediately after the price update or from the parts management report that is produced at the end of every month.

The posting has some pluses but one very big drawback. The pluses are similar to those listed above. The big minus is that the credit normally is posted to the department gross profit and thus increases the gross generated by the department. An example of the entry for GM is:

Parts Inventory	Acct 242	$1543.32+
Parts Cost of Sale	Acct 687	1543.32-

Note: **To record the appreciation in the parts inventory after loading the GM manufacturer's update date 9/1.** I don't believe that this price increase should be shown as an increase in gross because:

- It generates a fictitious "profit" because practically all the parts involved in the price increase may still be on the shelf and unsold.
- It generates a pay increase even though no transactions have taken place.
- It leads to a negative cash flow because it pays part of the "profit" away in bonuses.

Does this means that the price update should not be made? No, it should be made, BUT the credit should not go into profit. It should go into an account that I call "reserve for price increase." Set this up in the general ledger. Every time there is a price increase, the credit should go to this account. An example of this entry is:

Parts Inventory	Acct 242	$1543.32+
Reserve for Apprec	Acct 242R	1543.32-

Note: **To record the appreciation in the parts inventory after loading the GM manufacturer's update date 9/1.** This account should revert to a zero balance only after a physical inventory has been taken. To take this into profit after the inventory, there are a few steps to follow. For example, if the price increases for the year totaled $12,000 then this amount will appear as the balance in the reserve account. Let's assume that a physical inventory was then taken that resulted in a shortage of $8,000. This shortage would go against the $12,000 and leave a credit balance of

$4,000. This is the amount that should be credited to parts gross profit. An example of this entry is:

Reserve for Apprec.	Acct 242R	$12000.00+	
Parts Inventory	Acct 242		$ 8000.00-
Parts Cost of sales	Acct 687		4000.00-

Note: **To realize the appreciation in the parts inventory after physical inventory 2/1 and adjusting for a shortage of $8,000.** However, before doing this, I suggest that one more step be done. Create a printout that lists all parts that are non-returnable that haven't had a sale in 12 months. Scrap these parts and take the amount scrapped off this reserve as well. The balance left then should be credited to the parts gross profit. At this time, the parts manager would have the appreciation amount included in his bonus check. The parts manager's reward should depend on the integrity of the internal controls in the parts department and lack of obsolescence.

Cores

What exactly is a core? About 20 years ago, a core was simply the damaged part that was replaced in the shop. Parts departments would have a technician or local machine shop rebuild the core and then sell the core as a rebuilt part.
The parts department might give the service department a small credit for the damaged part and add this cost to the rebuilt cost. This gave the parts department a "like new" part to sell at about half the cost of a new part. It didn't take the factories and parts suppliers long to figure out that there was a big market in rebuilding parts, Since they wanted all those cores to go back to them instead of being rebuilt, they started charging a "core charge" anytime they sold a new part or a re-manufactured part that could be rebuilt and had a core value. They quit calling them rebuilt cores and renamed the parts "genuine re-manufactured parts". So basically a core in the parts inventory is one of four things; a "deposit charge" on a new part that is in stock; an "in transit" core that has been given credit to a customer for returning to the dealership and is waiting to be returned to the factory or parts supplier; a "work in process" core that the parts department is waiting for a technician to return; or a "damaged" part that the shop is rebuilding. Each type of core should have a different method of accounting for tracking them in the parts inventory.

Deposit Cores

This core type is easy to account for. The amount of the core charge is disclosed on each supplier's invoice and the core charge should go to a parts sub account. When the core is sold on a counter ticket or repair order, the sub account is relieved. They may need to change the computer setup to account for the sale of a core this way. It would be nice to put the cores in another parts source and create a pad for them, but currently the manufacturer update tapes only include core charges as a memo field on the inventory record of the corresponding part. You can get a listing of these cores, either though the parts management reports or by writing a special report, but you'll have to manually add them to the pad totals.

In Transit Cores

The second kind of core inventory is actually a receivable, but for simplicity we can make it a debit to an accounts payable account. Whenever they receive a returned core and issue the customer a credit memo either on a repair order or counter ticket, they should send the core back to the supplier and make an internal credit memo. The credit memo should have a preprinted number, the supplier's name, the core returned, quantity, price, and the parts counter ticket or repair order number. The office would post this credit memo in a special journal or source as a debit to the supplier's accounts payable account and credit the core inventory sub account. They should ask their supplier to use this credit memo number as the purchase order number on their credit memo invoice. The office would match the vendor's credit memos with the internally generated credit memos. I realize that this seems like an extra accounting step, but it will guarantee that they get all their core credits, If all cores have been returned to the suppliers, and all credit memos posted, then there will be no hand listing and the computer listing provided by the parts department will match the GL balance. They can now monitor this monthly instead of only at the annual physical inventory.

Work in Process Cores

These are the cores on repair orders. The dealership should have a firm policy that all cores get charged out on repair orders and counter tickets. When the technician returns the core, issue a credit. This may be an extra step, but this is what we have the computers for.

If the service department tries to charge a customer for the core and then has to stop everything and get the technician to bring back the core to parts and get the credit issued, they will slowly learn to return the cores before closing the repair orders. The same goes for counter tickets, no matter how well they know the customer, they should still charge them for the core and issue a credit when they bring it back. To see if this is happening, pull the counter tickets of some of the dealership's technicians and friends and see if the part that was sold had a core charge or not.

Cores being Rebuilt

This last type of cores shouldn't need any accounting for. According to most parts managers, this is not something that a dealership should be doing. Now that the price of re-manufactured parts is so competitive, rebuilding a core that has no warranty or return allowance can not only be risky, but enable the shop to "pad" hours on the core.

The only exception would be special line cars and heavy duty truck dealers that don't have access to re-manufactured parts. These cores should have a special part number, their own source, and GL account. They should be treated as any other NPN or accessory inventory. In summary, to find how many of each core type is in the inventory, run a computer listing of cores. These are the cores that are deposits on parts in stock or that cannot be returned yet. As long as the dealership is not rebuilding cores, the remaining value of core inventory is for cores that need to be returned.

Preventing Shortages and Obsolescence

Most of this course has mentioned various controls and processes to prevent shortages and obsolescence. In summary these are:

- Install a policy of internal controls and review them regularly.
- Analyze the ordering criteria to prevent carrying parts that will become obsolete.
- Install a special order parts system that identifies any parts not picked up before the factory's deadline to return expires.
- Encourage employee honesty by example.
- Make sure proper accounting procedures are being followed with regards to prices, returns, allowances, and discounts.

Quick Check

☑If you have a shortage in the physical parts inventory, check to see if the cost was adjusted on parts purchased from other dealers at a markup. Pull the accounts payable folder for a dealer and back track the invoices to the repair orders. Verify the cost amount or look for an adjustment on the invoice.

☑Make sure that no technician can remove a part from the parts department without it being charged out on a repair order. As a suggestion, have the parts department open a ticket for each technician and charge any “borrowed” parts to this ticket until the actual repair order number is provided.

Worksheets for Analysis of ratios in the Parts Department

GROSS TURNS = $\frac{\text{ANNUALIZED COST OF SALES}}{\text{INVENTORY VALUE}}$

Guide 4-5 times per year

TRUE TURNS = $\frac{\text{STOCK ORDER PURCHASES}}{\text{INVENTORY VALUE}}$

OR ...GROSS TURNS x STOCK ORDER PERFORMANCE

Guide 3-5 times per year

STOCK ORDER PERFORMANCE = $\frac{\textbf{STOCK ORDER PURCHASES}}{\textbf{TOTAL INVENTORY PURCHASES}}$

LEVEL OF SERVICE

	TOTAL GROSS SALES DEMAND
-	**EMERGENCY PURCHASES**
-	**<u>LOST SALES</u>**
=	**TOTAL NET SALES DEMAND**

$\frac{\textbf{TOTAL NET SALES DEMAND}}{\textbf{TOTAL GROSS SALES DEMAND}}$ **= LEVEL OF SERVICE**

The Accounting Office

The accounting office is often considered its own department in a dealership although they don't have any sales or expenses allocated to them.

The office has numerous daily tasks that include processing the sale of vehicles, service repairs, and parts, collecting funds and depositing them in the bank, paying the bills, and billing customers. Monthly, the accounting office does an accounting close and prepares a financial statement that is transmitted electronically to the manufacturer of the vehicles that the dealership sells.

In general the accounting office is concerned with the following areas;

- Asset Management
- Internal Controls
- Financial Reporting
- Expense Control
- Helping the other Managers

In this course, we will discuss these areas of concern and suggest methods to provide control and management over these topics. Before addressing these issues the next section will give you an overlook of the daily operation of the office.

Quick Check

☑Make sure that the goal of a quick monthly close and financial statements don't get in the way of helping the other departments retain profit.

DAILY OPERATIONS in the Accounting Office

A dealership sells three basic items; vehicles, service and parts, but one of the largest labor volumes in the accounting office is the "working of car deals." The vehicle sales are called car deals and the office processes these deals for three main reasons; to transfer the title of the vehicle to the customer, send the financing contract to the financial institution, and enter the sale into the sales journals. Dealerships normally do not own the vehicles that they have in inventory. The vehicles are floored at either a local bank or captive finance company owned by the manufacturer. When a dealership sells the vehicle and has collected the funds from the customer or financing institution that financed the sale for the customer, they payoff the flooring. If they do not have the funds to payoff the vehicle or they are delayed in paying off the flooring, it creates an "out of trust" situation with the bank. The bank normally inspects the vehicles each month during a "floor plan audit."

Car Deals

Some of the steps to work a car deal are as follows;

1. Check receipts for the down payment or full payment
2. If the deal was financed, pull the contract, credit application, and any other documents required by the financing institution.
3. "Bill out" the deal to enter the accounting entries into the sales journal
4. Submit or transmit documentation necessary to obtain the rebates or incentive from the manufacturer
5. Prepare information on trade-in to bring into it stock
6. Issue a check to payoff the customer's loan on their trade-in
7. Payoff the floor plan on the purchased vehicle
8. Prepare the titling information for the state licensing department
9. Figure the sales person's commission on the deal and prepare a commission voucher

Car Deal Posting

The following is an example of the accounting of a car deal. As you can see, some items affect the gross on the deal, some figure into the commissions paid to the salesperson, and some are used to come up with the amount financed.

Sales Price	$20,000	-$20,000	-$20,000	$20,000
Sales Tax	exempt			
License fees	exempt			
Flooring	$18,000	$18,000	$18,000	
Holdback	$500	-$500		
Roof rack	$250	$250	$250	
Trade in	$6,000	$6,000	$6,000	-$6,000
ACV	$5,000	-$5,000	-$5,000	
Payoff	$685			$685
Down payment	$1,500			-$1,500
Finance earnings	$480	-$480		
Extended Warranty	$745	-$745		$745
Cost Extended Warranty	$375	$375		
CLAH	$500			$500
Comm Rate	30%	-150		
Dealer rebate	$900	-$900		
Customer rebate	$750			-$750
Spin Cost	$25	$25	$25	
Pack	$200		$200	
Comm percentage	20%	**$ (3,125)**		**$13,680**
Payable Gross			**$ (525)**	
Commission @ 20%			-105	

The customer bought a new vehicle from the dealership for $20,000. There were no sales tax or license fees. The vehicle was floored for $18,000 with a holdback of $500. The dealership promised a roof rack at a cost of $250 to be installed later. He was given $6,000 for the trade in and its ACV was $5,000. The lien payoff was $685. His down payment was $1,500. The reserve on the finance contract was $480. He bought an extended warranty for $745 with a cost of $375. In addition, he obtained Credit Life and AH with a premium of $500 and a commission rate of 30%. There was a dealer incentive of $900 (not commissionable) and a customer rebate of 750 that went towards the down payment. There was a “spin to win” cost of $25 and a pack of $200. Commissions are 20% of payable gross. See if you come up with the same numbers as above for the Amount Financed, Gross Profit, and Commissions payable. The section on the Sales Department earlier in the manual might help you to understand the sales process better.

Parts & Service Sales

The accounting office does much less for the service and parts sales. Most dealership computer systems are integrated to these departments and the sales are posted automatically. There is an interface that determines which type of sales and payments goes to which GL account numbers. An edit listing is produced by the computer system and a clerk verifies the entries and updates the computer. Any repair orders or parts tickets that did not update properly are normally listed in an error report. These must be either manually posted or corrected and updated.

The Daily Deposit and Cash Sales

The main functions of the accounting office as it pertains to service and parts sales is to reconcile the cash collected to the sale documents. This clearing account is called cash sales clearing. Dealerships have numerous problems getting a clean reconciliation of this account. The cashier starts the day with a set amount of cash in her cash drawer. During the day, she collects funds for repair orders and sometimes parts tickets. It is best that she either keep a log of all her collections and balance that to the cash drawer or use a cash register tape. At the end of the day, she leaves the set amount in her cash drawer and turns in all the collected funds and documentation to the office. The cashier should not do the bank deposit herself. If the dealership is computerized, the office should receive a report that indicates which invoices tickets have been charged to cash clearing. This total should agree with the documentation provided by the cashier. Two deposits are normally done; one for the parts and service cash sales and another for the miscellaneous funds that the cashier collected (down payments, car payments, license fees, etc...) There should be a firm rule in the dealership – cash sales must equal zero at the end of the day. If there is a ticket left unpaid in the cash sale account – it wasn’t a cash sale – was it? Create a JV to credit cash sales for that ticket and debit “one time charge” accounts receivable – or make the parts or service department reverse the ticket.

Charge Sales

Charges in the service department are usually either fleet accounts or payments due from extended warranty companies. Charges in the parts department originate from sales to wholesale customers. The service sales are documented on a repair order, internal (for sales to other dealership departments) and warranty claim (for reimbursement from the manufacturer.) Hopefully the dealership puts customer pay, warranty, and internal repairs all on the same invoice. This streamlines the paperwork tremendously. Unfortunately, some computer systems require that all pay types be closed at once to complete the repair order. If a customer comes in and has warranty work done and some quick service like a lube, oil, and filter (LOF), the repair order cannot be closed until the warranty administrator finalizes the warranty portion of the repair order. This creates a timing problem with cash sales. The customer has paid for the repair order, but the accounting copies will not go through until a few days later when the warranty is closed. To correct this problem, some dealers put warranty and internal repairs on separate repair orders from cash sales. The best solution is to require that the warranty portion be closed immediately. This requires that the service advisor, technicians, dispatcher, and booker be careful to complete all the information that is required for a warranty such as trouble and failure codes. Another problem arises when the cashier closes a ticket but does not receive the payment. On some computer systems the cashier must close the ticket to get it ready for the customer to come in and pay. If the customer does not pick the vehicle up for a few days, the accounting entries show a closed repair order, but no cash payment. The same problem is created in the parts department. A parts counter person will create a parts ticket for a customer that is going to come in later to pick it up. Or the ticket must be created for the parts driver to take the part and deliver it to the vendor. Since these two types of customers are often cash only (charge account not set up,) then the ticket is set up as a cash sale. If the customer doesn't pay for the ticket for a few days, or never does, then cash sales account gets out of balance.

If there is a problem with cash sales created by these examples, then it may be easier to schedule cash sales for better control. Any open repair orders or credits can be investigated when they become overage. For more information on cash sales, see the section on Asset Management; "Reconciliation of Cash Sales." The service sales are recorded normally in four journals; cash and charge sales journals for repair orders, internal sales journal, and warranty claim journal. The parts department creates parts tickets. These are accounted for in a cash and charge journal also. Most dealerships do not have a good method to bill companies for the work that has been done. As a suggestion, the extended warranty billing should be on a separate accounts receivable schedule and given to the warranty administrator for billing and collections. The cashier should have a written policy for billing customers and all charges must be approved by the accounting office. Although this is usually controlled in the parts department, the same control is often lacking in service. Repair orders are left open and in the "bucket" until a customer pays or someone tells the cashier to write off the ticket. A daily listing should be made of the completed repair orders in the "bucket" and someone should note if the vehicle is still on the premises or if it has left the dealership. If it has left the dealership, then it has become a charge by the nature of what happened. A special schedule should be made for these "unauthorized charges" and the customers billed anyway. It is better to complete them to accounting where they can be collected and monitored instead of leaving them in the cashier's office as an open repair order.

Quick Check

☑Make sure that cash sales clearing account has either a zero balance every day.

Daily Operating Control (DOC) or MIS

After all the car deals have been entered and the service and parts journals have been updated, the daily sales are reported on a computer generated report called a DOC or MIS. DOC stands for Daily Operating Control. This report should be provided daily to the dealer, general manager, and department managers. Other functions that are done but sometimes not daily are payroll processing, accounts payable, and accounts receivable. Some of these functions are combined and done by the office manager or controller. These three functions are similar to other industries.

Month End

Monthly the dealership does an accounting month end close and prepares a financial statement. This statement is a preprinted form provided by the manufacturer with a common chart of account for all their dealers. Unfortunately, each manufacturer has their own chart of accounts and financial statement. If a dealership sells Toyota and Fords, then the accounting office needs to prepare two different financial statements. The last section of this manual includes a checklist for closing the month under "Computer Utilization."

Understanding the Flow of Accounting Data

Journals

In the previous section we discussed various documents in a dealership; car deals, repair orders, parts tickets. Other documents include checks, receipts, new vehicle purchase invoices, warranty credits, and journal vouchers. These documents are entered into the general ledger via journals or sources. An example of a listing of journals or sources is as follows;

Journals	**Description**
10	New Vehicle Sales
20	Used Vehicle Sales
30	Repair Order Sales
40	Payroll
50	Cash Receipts
55	Warranty Credits
60	Cash Disbursements
70	New Vehicle Purchases
75	Accounts Payable Vouchers

Each one of these journals or sources is summarized and flow into the general ledger. The general ledger is used to create a financial statement. Most dealerships use a financial statement format that is required by their manufacturer. It differs from the general ledger by listing sales and gross profit for each sale account instead of sales and cost of sales. Most sub accounts are grouped together on one line of the financial statement; for example the general ledger might have three accounts;

Sublet Inventory	$300
Gas, Oil, & Grease	$200
Paint & Body Materials	$250

The financial statement may group them as

Other Inventories	$750

In order to get a break down of large grouped accounts, you will need the trial balance or general ledger. If you want to know what makes up an individual account such as "Accounts Payable -Trade Creditors", you might need two items; the break down of each account on that line and a detailed schedule.

Schedules

Dealerships can often have over 100 scheduled accounts. Schedules have 3 basic formats; Balance Forward, Detail Forward, and Aged. Cash in Bank and Accounts Receivable are examples of Balance Forward schedules. When a new month starts, only the previous ending balance for each control number or account is brought forward. The purpose of these schedules is to provide detail during the month for information and reconciliation. A Detail Forward schedule would be New Car Inventory. All the detail for each control number stays in the schedule until the control number balance equals zero. Aged schedules are like Balance Forward and Detail Forward schedules except they have columns for Current, over 30, over 60, over 90, and 120+ days. Aged schedules usually have only one account on them, but Detail Forward can have 5 or 6 accounts "side-by-side" on the same schedule. An example of this would be a new vehicle inventory schedule that contains the new car inventory account, new truck inventory, demo inventory, and floor plan payable for each stock number. Some examples of schedules are:

Schedule Number	Name	Type
1	Cash in Bank	Balance Fwd
10	New Vehicle Inventory	Detail Fwd - side-by-side
20	Used Vehicle Inventory	Detail Fwd - aged
40	Warranty Claims	Detail Fwd - aged
45	Employee Receivable	Detail Fwd

The Balance Sheet

The balance sheet for a dealership is similar to the balance sheet for any other industry. It varies in that there is considerable about of detail for inventories and a break down of profit by each month. The following diagram is a simple balance sheet that contains three entries;

#1 The dealer invests $10,000 in the dealership.
#2 The parts department buys a part for $40 on account and sells it to a customer for $50 on account.
#3 The dealership buys a new vehicle for $20,000 and puts the amount due on floor plan. The vehicle is brought into inventory for $19,500 with a holdback due from the factory of $500. They owe the bank $20,000 for the amount of the draft.

The holdback of $500 is usually paid quarterly by the factory or when the vehicle is sold.

	Balance Sheet		
Assets		**Liabilities**	
Cash in Bank	$ 10,000	Accounts Payable	$ 40
Accounts Receivable	$ 50	Floor Plan Payable	$ 20,000
Due from Factory-Holdback	$ 500		
New Car Inventory	$ 19,500	Capital	$ 10,000
		Profit Year to Date	$ 10
Parts Inventory	$ -		
Total Assets	**$ 30,050**	Total Liabilites	**$ 30,050**

The Income Statement

The income statement comprises three or more pages of the financial statement. The sales and gross profit section details all sales by vehicle models, labor types, and part sale types. This section is normally on the last pages of the financial statement. The total sales and gross profit by department is brought forward to the top of the second and third pages. Below these totals are the expenses by each department. The expense format varies by each manufacturer. GM based statements have one long listing of expenses by each department and often an allocation of administrative expenses based on preset percentages. The Ford based statements make a subtotal by each department of gross profit less "selling expenses" to arrive at adjusting selling gross. Each department's adjusted selling gross is then brought forward to another page and administrative expenses are allocated. Most financial statements follow either of these general formats.

	New Car	Used	Service	Parts
Sales	**-20000**	-7000	-44	-23
Cost of Sales	18000	4500	14	18
Gross Profit	**-2000**	-2500	-30	-5
Percent	10%	36%	68%	22%
Commissions	400			
Advertising	25			
Floor plan	3			
Delivery Expense	2			
Salaries	600			
Training	5			
Policy	2			
Vacation	1			
Total Variable Expenses	1038			
Adjusted Selling Gross	-962			
Adm. Salaries	200			
Employee Benefits	30			
Payroll Taxes	70			
Taxes	40			
Insurance	300			
Telephone	45			
Data Processing	10			
Rent	200			
Total Fixed Expenses	895			
Net Profit	-67			

Detail by Model, C,W,I - Retail, Wholesales

Gross Profit Benchmarks

"Controllable Expenses"

3% Net profit to gross ratio

Understanding Sales and Gross Profit

If the parts department buys a part for $18 and sells it for $23, there is a gross profit of $5. The gross profit ratio is 22% ($5 divided by $23.) This is an important measurement in the service and parts department. They measure labor gross profit (usually 65-70%), parts on repair orders (35-45%), and parts sold over the counter (40%) and wholesale (25 %.) The sales department does not use gross profit percentages. Instead they use gross profit per unit ($1,500 per new vehicle.) Each item in a dealership has an inventory, cost of sales, and gross profit. There is no process costing or allocation of expenses to inventory costs except possibly when work is done to a vehicle or the parts inventory expense allocation done for income tax purposes.

MANAGEMENT in the Accounting Office

Employee Positions and Job Descriptions

The normal positions in the Accounting Department include:

Controller or Business Manager
Office Manager
Payroll Administrator
New Car Biller
Used Car Biller
Bookkeeper
Accounts Receivable Clerk
Accounts Payable Clerk
Title Clerk
Cashier
Warranty Administrator
Receptionist

Controller

The controller or business manager reports to the general manager, dealer, CFO, or in a factory owner dealership the managing division. In a factory owned dealership this position is called a Secretary/Treasurer. In a smaller dealership, this position is the office manager but the responsibility is normally similar. The controller is responsible for supervising the daily operations, preparing a monthly financial statement, and asset management. The objective of the controller is to provide a smooth flow of paperwork through the office, reduce expenses, and protect the assets of the company.

Office Manager

The office manager reports to the controller and assists the controller in the daily operations and supervision of the office employees. In some dealerships, this position might have responsibility for other functions such as payroll or accounts receivable. The office manager replaces the controller during vacations and sick leave.

Payroll Administrator

The payroll administrator reports to the controller or office manager. The position is responsible for the processing of payroll checks, maintaining the payroll files, administrating the employee benefit programs, and submitting the payroll tax reports.

New Car Biller

The new car biller reports to the controller or office manager. This person does all the functions listed in the previous chapter for new car deals. In addition, the new car biller often maintains the new and demonstrator inventory records. Sometimes this position is also responsible for submitting for rebate and incentive payments from the manufacturer.

Used Car Biller

The used car biller reports to the controller or office manager. This person does all the functions listed in the previous chapter for used car deals. In addition, the used car biller often maintains the inventory records for used vehicles. This position also processes the sales to wholesalers and purchases from auctions and wholesalers.

Bookkeeper

The bookkeeper reports to the controller or office manager. This is not a common position, but in some dealerships, the data processing of information is done by this position.

Cashier

The cashier reports to the service manager, assistant service manager, or controller. The cashier is responsible for service & parts customer payments, billings, and open repair orders. The cashier must provide a daily reconciliation of the repair orders and parts tickets that were paid for and the method of payment. The cashier balances the opening cash fund (drawer) to the ending drawer less payments received. The cashier is normally responsible for filing of repair repairs, collection of customer surveys, and processing of service reminders.

Warranty Administrator

The warranty administrator reports to the service manager, assistant service manager, or controller. It is the responsibility of the warranty administrator to apply for warranty reimbursement from the factory for repairs done on vehicles. The warranty administrator must follow the factory's warranty policy and procedure manual and attempt to claim the maximum amount of credit for the repairs. This includes careful attention to administrative and diagnostic time as well as the best price for parts and supplies. The warranty administrator maintains the warranty receivable schedule and must insure that claims are submitted and paid quickly. Any shortages must be investigated and reported to management.

Customer Relations Manager

This is a newer position to dealerships. During the past 5 years, the factory has required that the dealerships provide better customer satisfaction and retention. Many dealerships have added a QCP or CSI manager that acts as a liaison between the sales & service departments and the customer. The customer relations manager provides management with the results of a follow up call done usually 2-3 days after the customer picks up their car. The manager asks the customer if the work was done to their satisfaction and in the time frame expected. The customer relations manager attempts to satisfy any customers that are displeased with their service or sales experience.

Developing Long-Term Employees

One of the biggest personnel challenges that face a controller or office manager is how to keep good employees during the "slow" times. Most dealerships have high and low cycles throughout the year and from year-to-year. The key to successful management of the office is to have seasonal functions. If the times are slow, the billers can double as customer service representatives, helping to maintain the customer database. The other office staff members can weed out the long term storage for record deletion. This is a great time to provide training to the staff. The important thing is to manage the staff's time and provide projects for these slow periods.

Forecasting, Planning and Benchmarks - the Accounting Office

There are certain productivity goals that are regarded as benchmarks in the automotive industry. They are national, but in almost all instances, will apply to most dealerships.

- Office Employees to Dealership Employees 1 to 14
- (not counting warranty, cashier, or receptionist
- Billers to car deals 1 to 100
- Billers to F&I Managers 1 to 1
- Cashiers to Advisors 1 to 4
- Receptionists to phone lines 1 to 10

Encouraging Employee Honesty

Internal controls (discussed in a later section) are an essential management function to protect against theft. Another method is to encourage employee honesty through example. It is difficult to expect office employees to be honest about the supplies, postage, and photocopying when the dealer, other managers, or even the office manager takes these items without reimbursement to the dealership.

Providing Customer Service

Office employees should attend regular customer service training seminars. These employees spend the same amount of time on the phone and in person with customers as other employees. It must be remembered that customers are not just the people who fill out surveys, but are also the other employees, the sales department, and factory personnel. It is important to establish a good relationship with the other departments. The office depends on the sales, service, and parts departments for their staffing level, so they should treat them like customers.

Quick Check

☑To find out if there is a customer service problem in the accounting office, try a survey completed by technicians, parts counter people, sales people, and other managers.

ASSET MANAGEMENT

The Objective of Asset Management in a Dealership is:

- Establish an asset management policy
- Make goals for all current assets.
- Design a program to monitor these current assets against the goals
- Convert all idle assets into cash – cash flow.

Asset management can be as simple as to reduce the time from when the transaction takes place to when the payment is received. This is called time flow.

Creating your Asset Management Policy

Simple Policy of Time Flow

Car deals – into accounting within 2 days
Contract in Transit – 7 days
Vehicle Receivables – 12 days
Accounts Receivable – 30 days
Warranty Claims – 7 days

Balance Sheet Management

Cash On Hand:

Cash on hand is normally comprised of the petty cash fund and the cashier's drawers. A reasonable amount to have in this account depends on the size of the operation and a justification for the funds. The amount should be divisible by $20 or $50. In a normal dealership, you might have $200 in the day cashier's drawer and $50 in the night cashier's drawer. The cashier starts the day with a set amount of cash in her cash drawer. During the day, she collects funds for repair orders and sometimes parts tickets. It is best that she either keep a log of all her collections and balance that to the cash drawer or use a cash register tape. At the end of the day, she leaves the set amount in her cash drawer and turns in all the collected funds and documentation to the office.

Quick Check

☑Slip $20 into the cash drawer and see whether they have a credit appearing in this account the next day.

The cashier should not do the bank deposit herself. If the dealership is computerized, the office should receive a report that indicates which invoices tickets have been charged to cash clearing. This total should agree with the documentation provided by the cashier. Two deposits are normally done; one for the parts and service cash sales and another for the miscellaneous funds that the cashier collected (down payments, car payments, license fees, etc..)

Cash Sales

Transactions that would distort the cash on hand account are the items from cash sales that didn't clear. A cash sales balance arises in service when the cashier closes a ticket but does not receive the payment. On some computer systems the cashier must close the ticket to get it ready for the customer to come in and pay. If the customer does not pick the vehicle up for a few days, the accounting entries show a closed repair order, but no cash payment. It is also common that the parts department will sell a part to a customer that does not have a charge account and has not paid for the part yet. An example of this is a local gas station that calls and asks for the part to be held for them. The gas station does not have an account and the counterperson wants to have the invoice ready for the customer when they come in. The invoice is processed to the "cash sale" account and usually remains on the "will call" shelf with the part. Sometimes the invoice and part are sent out with the dealership's parts delivery driver. For control purposes, a copy of the invoice should always remain with accounting. Now that there are integrated parts invoicing computer systems, this invoice is often automatically downloaded into the accounting general ledger. This causes the cash sale account to have a debit balance. Trying to find which invoices are not paid is a difficult task, especially when this is compounded by the normal accounting errors of having a charge in the cash sale account and vice versa. If there is a problem with cash sales created by these examples, then it may be easier to schedule cash sales for better control. Any open repair orders or credits can be investigated when they become overage.

Methods of Cash Sales Reconciliation

Detail GL

If you detail the GL account for cash sales, then you can take the detail at the end of the month and "cross off" the clearing items with a highlighter. Each parts ticket and repair order will show as individual debit entries, but the cash receipt credit will be for the whole deposit. You will need some detail of this deposit which is normally stapled to the back of the cash receipt. By highlighting each paid ticket or repair order, you will be left with an amount open, which is hopefully the GL balance.

Scheduled Cash Sales

If the account is scheduled, then the outstanding tickets should be open items on the schedule. The control number is the repair order or parts ticket number and each receipt is posted for every cash sale item. Sometimes the cash sales are scheduled by the date and each control number should zero out.

Forced Zero Balance

This is the best method if the proper controls are in place. The account must be zero each day. All outstanding items are journal vouchered to a scheduled accounts receivable. These outstanding items are set up as "one time charges" with a control number that has no credit line. If necessary, a statement can be generated for these balances. There must be no exceptions to this zero balance; no "it will clear tomorrow" excuses.

Quick Check

☑If there is a problem with the daily balancing of cash sales, suggest that no checks be cashed from the cashier drawer until the problem is solved. This makes reconciliation easier.

In order to monitor this account more closely, cash sales clearing accounts should appear on the DOC or MIS report as separate line items. Then, at any time, the dealer and/or general manager can ask for a printout of the balance.

Cash In Bank:

Often the amount in the cash in bank account in not a valid number. Dealers like to show a healthy bank balance and will often "plug" the amount as a debit to one control number such as 9 or the month and credit accounts payable or notes payable. A key item to monitor is the dealership's average collected funds. This amount is usually on the bank statement and can provide information on how well the dealership manages excess cash. It is important to know how much is the collected funds balance required by the bank in order for them to waive the monthly fees. Collected funds are normally the balance in the bank available to pay checks from. This balance usually excludes out of state checks deposited that day or any other type of deposit that the bank can delay credit to collected funds by law.

Cash Management-The Daily Cash Report

Daily Cash Report

Cash per bank	$ 32,524.21
Cash per Computer	$(193,437.07)
Hold checks	$ 77,380.82
CIT submitted	$ 38,911.05
	$ (77,145.20)
Float	**$(109,669.41)**

Demands

Payroll 1-15	$ 73,250.00
Release payable 1-20	$ 32,528.25
Increases	
CIT and VR (less today)	$ 197,252.24

A Daily Cash report is a simple tool to provide to management your cash situation. The dealership should monitor its balance at the bank on a daily basis and get to understand the "float" or difference between the book balance and your bank balance. **** Click on Contact-Free stuff form at SandraJerome.com and click on the link for Super Controller zip file of spreadsheets.***

Cash Flow

Any excess cash should be transferred into an interest paying account. How do you determine how much money should be transferred? The answer is to use the simple cash flow management report that is shown on the next page. This cash flow report calculates what cash the dealership *should have generated* for the month. It takes the profit earned by the dealership for the month, plus or minus cash flow adjustments such as depreciation, the principal paid on a note, Federal and State taxes, and *elective increases or decreases* in assets. This is not a flow of funds statement that accountants prepare to show the changes in the structure of the financial statement from one period to the next. Rather, this is a report that shows how much cash they should have generated provided they kept a handle on all of the assets for the month. The transferred amount should be reflected as an increase in the general ledger account that has been set up in the D.O.C.

Cash Bonus Worksheet

	January	February
Balance Fwd	$0	$69,000
Net Profit	$120,000	
ADD:		
Depreciation	$10,000	
Holdback Release	$135,000	
CASH INFLOW	$265,000	
Less:		
Loan Repayments	($15,000)	
Capital Expenditure		
Holdback Increase		
Fed/State Taxes	($43,000)	
Elective Increase		
Net Cash Flow	$207,000	
Check	$138,000	
Balance	$69,000	
Bonus Percentage	1%	
Bonus	$1,380	

Please note that elective increases are amounts that have been *agreed upon before* the month starts. For example, if the dealership agreed to increase its used-vehicle inventory by $100,000 then that's the amount that should be entered on the form no matter how much the actual used-vehicle inventory has increased. If the used-vehicle manager exceeded this elective increase by $50,000, this would show in the days supply being too high, plus adversely reducing the cash in bank balance.

Bank Reconciliation:

The process of reconciling the bank account is probably the most time consuming task for a controller/office manager. The real problem is that the checking off of cleared items is painstakingly arduous, time-consuming and requires absolute concentration. When reconciliation normally does not pan out, the problem often lies in the listing of the uncleared or open checks. There are simpler ways of doing a bank reconciliation.

Here they are:

Method 1: (Electronic Spreadsheet Method):

Step 1:

Download the cash-in-bank schedule to a spreadsheet using communication software such as *ERALink or Reflection*. (They may instead use the computer's PC device.)
Step 2:
Using the same software package retrieve the cleared transactions from the bank and download this information into the same spreadsheet mentioned in *Step 3:* The download of all of this data to the spreadsheet will isolate uncleared or mismatched transactions. Clear up the mismatched items by using the VLOOKUP function in Excel, adjust for the miscellaneous charges and the balances left on the spreadsheet should be a listing of all open items. Certain dealers are doing this reconciliation every working day. It keeps the dealership current on its cash position as well as knowing exactly the status of its outstanding checks. This method is extremely fast and effective and requires the minimum of maintenance.

Method 2: (Bank produced reconciliation.)
Step 1:
Provide the bank - either via a modem or by sending the bank a computer disk - with the cash in bank schedule.
Step 2:
The bank will then compare its records of the account with the data that was sent by the dealership. Any open items or items which do not match the bank's records will be listed for them. This is in essence the open item listing that they need to undertake the bank reconciliation. This "bank assisted" reconciliation is offered by many banks throughout the US. The cost is reasonable and would save them money and time as well as strengthening the internal control procedures.

Method 4: (In-house Computer Applications R+R, CDK, Power R+R (UCS), DealerTrack, DealerStar, etc.): Most DMS computer companies offer a bank reconciliation program. This program prints out a list of outstanding items and a bank reconciliation.

Contracts In Transit:

Contracts in Transit are amounts due for financed vehicles. When a customer buys a vehicle on payments from a lender that the dealer has an agreement with, the amount due the dealer is paid directly to the dealer after the proper documentation has been received. Often these payments are in the form of a draft that the dealer prepares or an electronic funds deposit into their checking account. Other times the lender has a special bank account that they credit contract proceeds with and the dealer can write a check from this account to pay them the amount owed.

The amount of contracts in transit due can grow quickly after the end of the month rush and often not enough attention is paid to the balance since the payments are

usually a "sure thing." To do an analysis of the balance to find out if it is excessive, please complete the following table. It pertains to the accounting set-ups used in the dealership to track vehicle receivable transactions and days supply. If there are certain transactions that are not done by the dealership, please mark in N/A (Not applicable.)
This analysis can often provide a new insight into contracts in transit. Often the management will not be concerned with the schedule since no items are over 30 or 60 days old.

54	**Contracts in Transit**		
55			
56	Contracts in Transit	$177,585	YTD
57	Accounts Rec. - Vehicles	$205,252	YTD
58	Total Receivables	$382,837	Line 56+57
59			
60	Avg Amt. Financed	$ 9,242	Input
61			
62			
63	No. of Retail Sales	143.0	Input or unit count
64	Sales per Day	5	Line 63/30 (days)
65			
66	Actual Days Supply	9	Line 58/60/64
67	Objective	3	Input - guide
68			
69	Excess in Days	6	Line 66-67
70	Dollar Excess	($250,676)	(Line 68*60*64)

But what if the dealer is carrying a 20 day supply of contracts in transit when the average payment days are only 3 or 4? Some of the ways that the payments can be accelerated are:

- Send contracts to the banks by courier or overnight mail
- Have the finance managers pull the contracts and send them to the bank
- Verify exactly what documents the bank needs to buy the loan. Dealerships often send in more than what is required.

Service & Parts Receivables

Dealerships should have few receivables unless they have a large wholesale parts operation. Most service customers should pay by cash, check, or charge cards. If the dealership is in the habit of setting up a charge account for good service customers, discourage this. It is better for the customer and the dealership to have the customer use their charge cards. Not only will the customer earn frequent flyer or other bonus points, but the dealership will avoid collection problems and possible damage to customer retention.

At what stage - in days - would they regard a service and parts receivable account to be overdue - 30, 60, 90 days? Just as in the contract in transit analysis done in the previous section, the days supply of P&S receivables should be calculated. If you look at a summary GL for the month (a summary GL, shows the total charges by journal) for the accounts receivable account, you will see how much was charged. Complete the following:

This spreadsheet (free downloads) calculates the days supply of P & S receivables:

22	**Parts and Service Receivable**		
23			
24	Approx. Receivable Sales:		
25	Body Shop -all AR		
26	Labor Cars	$25,212	MTD
27	Labor Trucks	$12,525	MTD
28	P&A Cars	$15,252	MTD
29	P&A Trucks	$13,252	MTD
30	Paint Material	$42,325	MTD
31	Sublet	$1,252	MTD
32	Parts Department		
33	Wholesale Parts	$1,232	MTD
34	Total Receivable Sales	$111,050	Total lines 26,27,28,29,
35			30,31,33
36	Total P&S Receivable	$195,252	YTD
37			
38	Variance (Excess)	($84,202)	line 34-36

Accounts Receivable Management

If possible, separate the receivables into parts, service and body shop receivables for management review. This can be done by using a separate GL account for each type of account. Another method is to assign customer numbers with a prefix for the type of customer. If you are using the phone number for the customer number, prefix it with a "P" for parts accounts, an "S" for service accounts, and "B" for body shop accounts. Since the schedule prints in control number order, the B's will print first, then the P's, and finally the S's. This segmentation only works if your customers are defined by department. If you are in a marketing area where the customers are both service and parts customers, then strive for a common account and one statement. The detail schedule will tell you by journal source which past due amounts belong to which department.

Each week the department managers and the controller and general manager should have an accounts receivable meeting. It should be a short meeting to review those accounts that are past due. Since departmental managers' pay is affected by bad debts charged to their departments, they should be responsible for collection efforts. Their employees are in direct contact with the customers and know the situation better than an office clerk that is just calling a name and number. If the accounting office becomes responsible for collections and reporting past due accounts, eventually one function will suffer.

NSF Checks:

NSF checks need to be immediately accounted for. The best method is to have a NSF account receivable schedule that produces statements. As soon as a NSF check is received, check to see if maybe the funds are good and can be redeposited. If so, make the entry to credit cash in bank and debit the AR schedule. Make another receipt to "redeposit NSF" check and include the check in the deposit. If the funds are not good, make the entry to credit cash and debit the AR and make 3 copies of the NSF check. One copy goes to the sales person, counterperson, or service advisor. The 2nd copy goes to the department manager, and the 3rd copy to the general manager. Putting the responsibility to collect the NSF check into the hands of the employees who know the customer and are paid on the profit from the sale helps the collection process. The main function of the accounting office is to report the collection requirements. When this responsibility is also combined with the collection responsibility, the reporting accuracy is less. Human nature is to avoid reporting bad news that they are responsible for. In order to notify all employees of a NSF check, change the customer's name in the customer database from John Doe to John Doe-NSF. When this customer comes in for service or parts, the name will clue employees to notify the departmental manager.

Other Receivables

These other receivables can be many things, sometimes a "plug" for profit or out of trust situations or valid receivables such as employee or owner notes. Each type of receivable should have a sub account or be scheduled.

Inter-Company Transactions

These are one of the most difficult accounts to reconcile in a dealership group. To solve this problem, join all the chart of accounts and centralize the accounting. This may not be an option for the dealership group, so a system that can be installed - is an old solution. No dealership may make a charge to an inter-company account without a purchase order number from the controller or office manager of the charged dealership. This settles all disputes. Another option is that only one company can make inter-company charges. One dealership is selected as the lead dealership and all expenses that are to be spread between the dealerships must come from that dealership and signed only by the group CFO or President.

Factory Receivables:

The contents of this receivable vary between the manufacturers. In general, the most important - and most challenging - factory receivable that needs managing is the incentive receivable money that is due the dealership. If the factory pays rebates every week, then the days supply at the end of the month should only be 1 week.

Finance Reserves

There should be about 30 days supply of Finance Reserves. There are a few reasons why the supply would be more than 30 days worth:

1. The finance company pays only quarterly.
2. There are adjustments (shortages) in the schedule that have never been adjusted.
3. The chargebacks have not been accrued.
4. A big push of deals at month end.

87	**Finance Receivable**		
88			
89	Finance Income New	$ 19,767	MTD
90	Finance Income Used	$ 6,956	MTD
91	Total Finance Income	$ 26,723	Line 89+90
92			
93	Days Supply Objective	$ 30	Input
94			
95	Objective Finance Receivable	$ 26,723	Line 91*93*3
96			
97	Actual Finance Receivable	$ 39,419	YTD
98			
99	Variance (Excess)	($12,696)	Line 95-97

Reconciliation Process for Finance Reserves

Many dealerships are still using the manual method of printing a detail monthly transaction list for the finance reserve and then manually writing the amount credited by the finance company next to each item. Then the difference is manually written next to each item where a difference exits. The differences are then totaled and adjusted to the finance reserve and the finance income.

This is a laborious process and should be automated. Set up the finance receivables as a schedule using a detail forward format. If possible, use the last name or the first four letters of the last name as the control number.

When the finance reserve statement is received from the finance company process this document as a source document through accounting, crediting the finance income amount to this schedule, the finance chargebacks to the expense account and the balance to a self clearing control number. (For example, they could use for the control number, the year + 01 for January, the year + 02 for February and so on.)

A special finance reserve journal is established in the computer system. The only two accounts that should go into this journal would be the finance reserve and the finance chargebacks entries. The total of all the entries made should equal the amount of the check from the finance company. Debit this check to the self clearing control which would now clear this balance.
The result of all of these steps would be a schedule that would easily isolate all outstanding items. They can set up the computer system to use the automatic write-off feature and write-off amounts that are insignificant.

Avoiding Chargebacks For Pay Plan Calculation Purposes

Some dealers who have chosen not to sign up for the “no chargeback option” from their finance companies, instead are reserving a fixed percentage each month of their finance income against possible chargebacks. This is done to smooth out the fluctuating chargebacks that they get and to gain credibility with their F & I department that a fixed amount of what is earned will be used to pay commission. At the year-end the reserve, if the appropriate percentage was used, should be close to “a wash.” Review this reserve for chargebacks account throughout the year in order to ensure the amount being set aside is in line with the actual chargebacks taking place.

Warranty Receivables

These are the amounts due from the factory for repairs made to vehicles under warranty. The warranty repair orders are often a part of the customer or internal repair orders and delays in submissions could be caused by a delay in closing the other portion of the repairs. Other delays are caused by improper submission to the factory or waiting for an approval from a district manager. The schedule may also contain warranty payment shortages or denied claims.

Calculating the days supply of warranty receivables:

6	**Warranty Receivable**		
7			
8	Warranty Labor Sales	$9,325	MTD
9	Warranty Parts Sales	$7,904	MTD
10	Body Shop		
11	Warranty Body Labor	$32,525	MTD
12	Warranty Body Supplies	$42,325	MTD
13	(materials or sublet)		
			Total lines
14	Total Warranty Sales	$92,079	8,9,11,12
15	Factor used	27.5%	Input
16	Warranty Objective	$25,322	line 14 * 15
17			
18	Warranty Receivable	$94,242	YTD
19			

20	Variance (Excess)	($68,920)	line 16-18

Internal Controls - Warranty Receivables:

A few safeguards should be in place in regard to warranty receivables;

1. The warranty write-offs should be approved by the service manager and controller if over a set amount (more than $1.00 for example.)
2. The warranty credit memo should be applied to the schedule exactly as paid. Any adjustments should be made by journal voucher.
3. Resubmitted warranty claims should retain their original date and hopefully the same repair order number.

Accounts Receivable - Officers and Employees:

Unless the dealership has a working capital problem with the manufacturer, this account should be used for all loans and advances made to officers and employees. In many dealerships these advances and loans are "buried" in the accounts receivable section of the balance sheet and get lost. The best way to monitor this asset is to have it shown as a separate line item on the financial statement. This normally leads to questions being asked and a reduction of this asset - unless it gets hidden away somewhere else.

LIABILITIES AND EXPENSES

Notes Payable - Vehicles

Dealers normally do not own the vehicles that they have on the lot for sale. It is common to own most of the used vehicles, but the new vehicles are usually "floored" with a floor plan lender. The dealer must arrange a floor line with a bank or captive lender when he becomes a franchised dealer and the agreement normally allows the factory to ship vehicles directly to the dealer and "draft" on the floor line. A captive lender is GMAC, Ford Credit, Chrysler Financial or any financial institution that is owned or controlled by the factory. When the draft arrives at the bank or captive lender, they put the vehicle by serial number on the floor line and pay the factory. Once a month (or other agreed upon time interval,) the lender or their representative comes out to the dealership and does a floor plan inspection. The dealer does not get prior notice of the visit. The lender physically inspects each vehicle on the floor plan report and requests payment or an explanation for any that have been sold or are missing. The dealer is judged based on this inspection and a good floor plan inspection can mean easier borrowing terms in the future. Once a month, the bank or captive lender sends an interest statement to the dealer. This statement lists each vehicle by serial number and shows the balance due, interest paid to date, and current month's interest charges. The total current charges must be paid in 5 or 10 days. This is usually one of the largest expenses after personnel and advertising that a dealership has. There is usually a separate expense account for this expense that shows in the variable section and should not be grouped into the "other interest" account that is in the fixed section of the financial statement. Periodically, this statement should be reconciled to the floor plan schedule at the dealerships. Some computer systems allow you to sort this schedule by the serial number and that makes reconciliation much easier since the floor plan statement is in serial number order and the schedule is in stock number order.

Example of a Floor Plan Reconciliation

New Vehicle Flooring per GMAC	$15,347,343.23
Less units paid off - after GMAC cutoff date (sch D)	-124,123.43
Plus units on dealer's schedule not yet floored (sch A)	137,343.23
Less units not belonging to dealer (sch B) -charged in error by GMAC	-24,123.42
Total Adjusted New Vehicle Flooring per GMAC	$15,336,439.61
New Vehicle Flooring Schedule - Acct 310	$15,147,333.23
Plus units in transit - not on dealer's schedule yet (sch C)	189,106.38
Total Adjusted New Vehicle Flooring per Dealer Schedule	$15,336,439.61

Accounts Payable

The accounts payable in a dealership comprised normally three things;

1. Trade creditors - purchases from other dealers and local part stores for parts and sublet labor.
2. Accrued expenses - accruals for utility, phone, advertising that hasn't been billed yet.
3. Factory statement - large account that has all the transactions between the dealer for each manufacturer; part purchases, expenses such as brochures, programs, and signs, and reimbursements for rebates, interest protection, and warranty claims. This must have a monthly reconciliation done for the balance to the manufacturer statement.

The trade creditors normally bill by statement and each invoice is posted individually into a control number for that vendor. Dealerships are either on open item accounts payable or balance forward. If they are on open item, then each invoice is approved for payment, and a check is issued automatically that clears each invoice that is paid. If they are on balance forward, then a reconciliation of paid and open invoices must be made by vendor. Either way is fine if it works for the dealer. The problem arises when they have all three types in the same schedule. Vendor types 1 and 2 are easier on balance forward accounts payable since they bill and pay by statement balance. Vendor type 3 - is easy to be on open item since many items do not clear immediately. The best solution is to keep the factory vendors in a separate GL and schedule. This schedule should be open item if handled properly. The other types of vendors can be either, depending on the preference of the office. Since open item accounts payable is a newer concept, it is often difficult for dealerships to adapt to paying by invoice. Instead, they continue to pay by statement and the items on the schedule never clears. If this has happened, a source needs to be added to clear items and clean up the schedule. The method either needs to be changed back to the way the accounts payable clerk is familiar with or training needs to be done on how to perform an open item accounts payable payment cycle.

Extended Warranty and Credit Life

When a customer buys a vehicle in the sales department, they normally have the paperwork completed by a Finance & Insurance Manager (Business Manager.) This person prepares the contract and also tries to sell a warranty that goes beyond the manufacturer's warranty in coverage and time. In addition, the manager tries to sell credit life and disability insurance that will either payoff the vehicle or make the payments if the customer dies or becomes disabled. The F&I manager may also sell aftermarket items such as alarms, rustproofing, or sunroof. The accounting for each of these is different;

3. Credit Life - has a premium for the term of the policy. The life portion has different commission rate for single or joint life and disability. A premium of $300.00 may be $100.00 for single life at a commission of 15% and disability of $200 at a commission of 20%. The total commission would be $55 ($15 plus $40.) Another premium of $300.00 could be $50 for joint life with a rate of 20% and disability of $250.00 with a rate of 10%. The commission would be $35 ($10 plus $25.) Normally the accounting is for the commission to go to an income account and the premium less commission goes to an insurance payable account. Since it would be a problem if the dealer collected the premiums and never forwarded them to the insurance company, this account should be scheduled by customer number and aged.
4. Extended Warranty - this is handled differently. The sale amount usually can be any amount determined by the F&I manager and the cost is determined by a rate chart issued by the extended warranty company based on model, year, and coverage. The cost of the extended warranty is treated as a cost of sale, and rather than a corresponding inventory credit, the credit goes to a payable

account. To ease the payment of the extended warranty statement, the credit is normally scheduled by the last 8 of the VIN# if the company bills that way. It is important that a customer name is entered into either the description field or control number 2 to make reconciliation of the schedule easier. This should also be aged to watch for older premiums that had not been forwarded.

5. Aftermarket - these are similar to the extended warranty sales except the cost of sale entry usually credits the vehicle inventory account instead of a payable. The sale account is sometimes the same as the vehicle sale account and makes reporting on these sales difficult. The best method of accounting would be to have a sub account by each model for aftermarket sales. The credit would go to a scheduled due bill payable account instead of the inventory. A detail of these sub accounts would provide the accounting necessary to monitor these sales.

Quick Checks:

☑These insurance and warranty schedules to monitor. The company can deny claims if a policy was not submitted in time. It is easy to pay for policies for customers that never bought the vehicle (unwinds or reversals.) There are many errors in costing the policies due to rate chart changes. In addition, there are refunds for cancellations that are waiting for reimbursements on the schedule.

Accruing Payroll

Dealerships are on a monthly accounting cycle and some payrolls are paid weekly. If this is the case, then the unpaid portion of the last week must be accrued. The easiest method for this is to take a weekly payroll and divide by 5 and multiply by the numbers of day to accrue. Only accrue the expense accounts and reverse the entry the next month.

Quick Checks:

☑ Some dealers run into problems with the work in process account due to this accrual. The solution is to use a clearing account for work in process. Debits to work in process are made based on hours booked. Credits are made to a technician flag payable account. This account is used for payroll, accruals, and reconciliation.

Payroll Taxes

The amount of accrual for payroll taxes comes from the dealer's computer system or payroll service. This account can be understated due to a lack of adjustment of the accrual rate. Often the state unemployment rate changes during the year from let's say 3.5% to 3.8%. This change is not made on the computer system, thus future accruals are under expensed. Another problem is the wrong cutoff point for wage base. Let's say FICA is taxable to $65,000 this year, but last years base of $60,000 is still in the computer. An under accrual will happen because it will stop accruing the dealer's portion of the tax at $60,000 instead of $65,000. An overstatement happens when the opposite happens; the rate decreases or the base decreases. Sometimes the amount is only an estimate and both situations could be happening.

Expense Controls

The most successful approach to controlling expenses is to have a process in place to screen the expense, the vendor and the prices being charged for the service. In addition, the dealership must have access to all expense information by producing a monthly expense report that trends each expense category and compares the expense level this month with the expense level of one year ago. To do this successfully, the accounting office must keep a clean and accurate set of books as well as prorate expenses uniformly and consistently.

Approved Vendor List:

Once a year review all vendors and select the approved vendors for the year. These vendors would be limited to supplying only the items that were approved.
Create a listing of approved vendors and refuse to sign checks for vendor purchases not on the list. This will reduce the expense exposure and the problem of every new manager bringing in their own retinue of vendors that they deal with.

Expense Reporting

If the dealership computer system provides it, request a full detail of expense accounts. All expense accounts should be either in a detailed GL (most systems will let the dealership decide which ones are detailed) or allow a custom report to be generated. The key to making this report useful is to require descriptions on entries. If the computer system allows it, the user can be forced to enter a description. This option is usually in the GL setup or journal setup. If the computer system does not provide a detail report, select certain accounts that need detail and either balance forward schedule them or create a book to enter the detail as it happens. This makes year end documentation much easier if the selected accounts are detailed monthly.

Trending and Ratios

Another key to expense control is trending the expenses. Having a computer report that shows how much each expense has been per month enables a comparison for variations. To go one step further, a report that shows a % of gross profit is helpful. Certain expenses such as sales commissions need to be monitored by this percent. If the normal sales commissions are 23% of gross, the amounts may fluctuate greatly between months, but the rate should stay constant. Other amounts, such as utilities, do not have a constant % to gross and should be monitored for any variation in the amounts.

DMS Computer Utilization

There are many systems in the dealership today, but the "core" system is the DMS or Dealership Management System. This system computerizes the accounting, parts, sales, service and parts departments. The most common systems – or traditional systems are CDK, R+R, and DealerTrack. There are actually over DMS 20 systems available today – maybe try an "alternative systems."

Training and Support

A lack of training is the cause of most computer system failure (other than "bugs" and hardware issues.) Dealerships are not like large corporations. When employees are hired, they are not sent to a training department to learn about company procedures and to attend computer classes. They must learn by trial and error. Some of the computer systems have interactive training available, either on the actual system or from a stand alone PC. Many dealerships do not realize that this training is available for new or rusty employees. A few of the computer companies still provide off-site training classes, but the trend lately has been to delete these and provide training only when a new system is purchased. As a solution, the dealership should invest in a system administrator that is responsible for training new employees and maintenance of the system. When a new employee is hired, the system administrator can review the procedures that are unique to that position or dealership and if necessary, fully train the employee on the system if they have never used that particular system before.

Easy Introduction to Report Basics

The true utilization of a computer system comes from being able to get out of the computer what you put in. The ability to create and run reports that the dealership can use, is one of the most important features. Most DMS systems have three types of reports;

DOC or MIS

The MIS or DOC -- Daily Operating Control - this user definable report contains 3 columns for GL balances - Day, MTD, and YTD and 3 columns for Forecast amounts. Each line can contain a single GL account, multiple GL accounts, other line amounts, or a fixed amount. The lines can also perform mathematical functions such as subtracting one line from another, or taking a percent of a GL account. The format of the columns (Day, MTD, YTD) cannot be changed, nor can the basic format of columns and lines. More complicated financial reporting requires the Custom Financial Statement.

*System Reports**

Each area of the system; accounting, service, parts, and sales (F&I and vehicle management) has numerous preset reports. Examples of these are schedules and journals in accounting, aged inventory analysis in sales, and open repair order listings in service. Often the prompts before running the reports grant the user flexibility in what will be contained in the report and the format. Some reports can even be customized such as many of the parts reports.

- *DMS systems change often, so check your DMS help guide for the latest reports

Report Generators

These reporting tools are based on a database concept. All information in the computer is stored in various files like a customer file, a GL balance file, or inventory file. The report can only use one file at a time, but some DMS systems are usually willing to create new dictionary items in a file to store information that is normally only in another file. The Report Generator uses normal database reporting concepts; you select the file, the sort, and the selection criteria for information, the output requirements, and format. This is an easy tool to use since it displays all the dictionary items (fields) while in the input screen. The basic requirements of creating a report are:

Select the File Name – where the information is stored. For example with CDK the car deals are in a file called FI-WIP.

Selection Criteria - What car deals do you want out of the big car deal file (all deals for a certain model, deals written this month, all deals for a certain salesperson?)

Sort - What order to do want to see them in (by stock#, by VIN#, by days in stock?) I normally recommend that you do not use the DMS system to sort, bring the data into Excel and sort much faster.

Output Fields - What information do you need on the report(color, stock#, desc, VIN#, date?)

I like to start with output fields - **select 4** first to see what is in the file and in what format. If you don't look at the file first, you might enter the date as 01/04/05 instead of 01Jan05 and get no deals on your report. By selecting the output fields first and running the report to the screen, you can see which are good fields and the format of the information.

Understanding the "AND" and "OR"

When you enter more than one selection criteria, the system will ask if it is **AND** or **OR.** These are two examples of the "and" and "or" prompt.

Sales date less than 1/1/18 **AND** greater than 12/15/15.

model number P53 **OR** P54

An **"AND"** example is a range of dates; greater than a certain date and less than a certain date. Both of these conditions must exist. An **"OR"** example is if either item would be okay. Let's say that the rebate program is good on models P53 or P54. This would be an **OR.** Sometimes the report writer doesn't like using multiple ands and ors and the report does not exclude what you want.

Looking at the F&I, Service, Parts Setups and Defaults

There are numerous items that control F&I, Service, Parts and their interface with accounting. For example, if a customer is a certain sales type, they don't get charged sales tax and their sales go to a special sales account number. Pay close attention to these setups which are usually in each module instead of accounting. If things are not coming into accounting the way you want, this is normally the problem.

Closing the Books in 3 days Using the Computer System

In some ways it was easier to close the books before computers. There was no delay waiting for schedules or reports to print and no need to purify the manual record to the computer record. The problem that has arisen since the reliance on the computer is the duplication of work. Why are the schedules compared to a hand generated wash out card? Why are schedules purified during the month and again during the month end close? The following is a simplified checklist for closing the books. The order works best with a highly computerized office and should produce a financial statement in 3-5 days depending on the cutoff day. To achieve a 3 day close, the deals must be in by noon of the first working day.

Before the last day:

___Purify all schedules except the vehicle schedules.

This will be the final purification for all schedules other than the vehicle schedules. The vehicle schedules should also be cleaned up, but the other schedules must have their final purification done before the end of the month. As a suggestion, use the following guideline:

Accounts Receivable *- purify before they send out the statements on the 25th.*

Accounts Payable *- after the statements have been paid on the 10th and the 25th, purify.*

Sublet Repairs *- same as accounts payable.*

Warranty Claims *- purify after each credit memo is applied and keep the final purification after last credit memo is processed for the month.*

___Make all Standard Entries

On the Last Day:

___On the last day of the month, make a list of all deals not in the accounting office.

1st thru 5th of the Month

___Provide the Sales Department with a list of all deals not in the accounting office at 9:00 am of the cutoff day.

___Close Parts & Service.

___Post final day parts and service transactions.
___Apply any remaining warranty credits that have been received.
___Produce an open RO and Parts Ticket report for managers.
___Clear/reconcile cash sales clearing accounts.
___Provide a list of policy expenses to the service manager and other expense detail that managers may want to see.
___Furnish parts and service managers with a finalized sales DOC.
___Post all Accounts Payable to the Inventory Accounts.

This is an important step. Do not let the accounts payable clerk go through the accounts payable vendors in alpha order - or post any of the large parts vendors. Post only the items that get added to the vehicle inventory accounts; tires, trim, air conditioners, detail, etc. When those are done, then continue with the other vendors.

___Cut off Car Deals at Noon of the cutoff day.
___Reconcile vehicle deals in the office with the deals in the F & I department.
___Post all car deals by noon of the 2nd day after cutoff.

This is a tough step. It requires that someone help the biller or billers to post the deals into accounting; possibly the office manager or the controller. Nobody leaves for lunch until all the deals are posted.

___Enter the remaining Accounts Payable invoices.

The first ones to be processed should be expense payables. Do not allow any invoices to be posted in the afternoon to the vehicle inventory accounts. These schedules are closed. It is not necessary to post the factory parts statement - just the invoices. This is just another vender and it should be accrued prior to the closed-not reconciled.

___Post Final Payroll.

If they must wait for the final payroll to close, post it now. All the commission vouchers should have been already been pulled from these deals. Sales commissions should be accrued anyway and do not affect the bottom line but speeds up the payroll process. Do not be concerned with commissions for deals that are going to be held over until the following month.

___Print schedules at lunch.

Only the vehicle inventory, contracts in transit, and vehicle receivable remain to be purified. The remaining part of the afternoon is spent working these schedules. The billers should do the vehicle inventories and the office manager should do the contracts in transit and vehicle receivable accounts. All should be fairly clean as these schedules were looked at and worked before the month-end.

___Obtain the flooring interest charge for the month and accrue.
___Post the remaining general journal vouchers.
___Run a DOC and accrue bonuses.

The DOC should have every line item that they need to accrue the bonuses and have an final income statement that reflects the data in the Operating Report. Compare the DOC to last month's - or run a comparative to check that all expenses are accrued and everything looks reasonable. If they currently do not have a functional DOC take the time to create a useful DOC as this will really help they with accruing bonuses and doing comparative analysis.

___Post final bonus general journal vouchers.

Run a tape of the bonus expense and then post the journal voucher. Verify the tape amount to the final DOC. If they want to make absolutely certain of the figures run a preliminary financial statement and verify again. This figure can now confidently be given to the dealer as the final number. The heat is off!

___At lunch print mini-schedules and reports that they need for additional financial statement information such as aging amount or memo counts.
___Obtain from payroll personnel counts for every department. *They may want to give the payroll clerk a copy of the personnel count section of the financial statement and request that it be completed by this time for inclusion in the posting of the memo accounts.*
___Make Memo entries and print the Financial Statement.
___Print the financial statement on blank paper and photocopy onto overlays. It is really unnecessary to provide original copies to anyone. If they have laser printers consider the new financial statement print software available from the computer companies.
___Make up backups and print necessary "hard copy" information at night.

Computer Backups

It is strongly suggested that a monthly backup be made for the accounting department. This backup should be kept in a safe place, possibly away from the dealership. Some companies like KoData and CD-One View take the backup tape and put it on CD. This is a good practice. Should the dealership lose their hard drive or whole computer in a fire, then some of the information could be retrieved from the backup. A daily backup is also helpful. Again, the controller or system administrator should remove the daily backup from the dealership and rotate in the backup from the previous days.

Super Controller 2 – Expenses and Employees

My Super Controller Guide Book 1 above, introduced office managers and controllers to the rest of the dealership. Hopefully you learned how each of the other departments function and how you can help others do their jobs better. The key to being a Super Controller is being in control of the whole dealership instead of just the accounting the office. The **Super Controller 2 - Expense and Employee Guide** takes the controller back into the office to tackle two problem areas; **Expense Control** and **Employees** but with an objective to reduce the expenses of the whole dealerships. Half of this new guide is spent on a different way to control your expenses and the other half is a guide on staffing and managing an accounting office along with creating pay plans and job descriptions for the full dealership.

In the **Expense Control** section you'll learn how to;

- ❒ Use technology like Excel to find out how much is being charged to an account
- ❒ Use my "item" method of attacking expenses
- ❒ Decide what should be in each account and create a metric
- ❒ Make a plan and budget to reduce expenses.

In the **Employee** section you'll learn how to;

- ❒ Create job descriptions using the samples I've included
- ❒ Develop pay plans that work
- ❒ Use tools like reviews and bonuses to get the results you want
- ❒ Determine the staffing of your office or centralized accounting department

As always, my "take" on these two topics is a little different as most controllers, office managers, CPAs, or consultants. Hopefully you'll get some ideas to help you in these two areas – they were always a challenge for me, but over the years of being a controller and a consultant, I discovered some ways to make the job easier and get better results. Read on!

Part 1 - Expense Control

I used to hate working on expenses. Not only did it take days to detail out the account, as a controller I ended up feeling responsible for expenses being high for the month. I would try my best to cut expenses in the office but in reality it is pretty hard to straighten out staples and reuse them. You just end up jamming a stapler. That brought me to the realization that my job is to not only control the expenses of my accounting office but to start working on the big expense accounts by analyzing expenses and helping the departmental managers reduce them. Yes, if I order file folders from Staples instead of a computer vendor, I'll save a few thousand a year – but if I can help create processes that reduce the amount of days it takes to get a vehicle to the front line and sold – I can save thousands per month. If I can make sure internals get posted to vehicles, contracts get pulled sooner, and reduce overpayments of commissions – I can save thousands per day!

If you start looking at expenses as a game – a challenge – a mystery to be solved, it will be a lot more fun that the old drudgery of "breaking out expenses" and trying to explain why they went up last month. If you figure out how to reduce expenses – you'll be a hero. For those of you that are new to our industry – our "after the fact" method of expense control is a mystery to you. We rarely budget in our industry – and even though we might create a forecast – it ends up being more of guessing game. Before we get started, you might ask – "why all this focus on expenses?" Can't we just gross profit ourselves into a better month? The answer is that expenses rarely go down. Even when we have a slow month and sell fewer vehicles, floor plan expense and advertising per unit to goes up. Gross profit is low, and we paid more minis causing the sales commission variable expense percentage to increase. To complicate matters, towards the end of the month, the sales managers went into a panic and overspent the advertising budget in the fear of having a bad month. It ends up being a self-fulfilling prophecy – the increase in expenses coupled with the lower gross profit caused a double whammy and ouch – a bad month! How can you get your expenses under control so you won't have such a hill to climb? You need to create a budget that is not based on units or gross profit – but reality. This budget process includes discovering what is being charged to each account and why. Before you can start creating a budgeting and processes, first you need to know what is in each account.

Detailing the Expense Account

First try to use a "base account: concept for detailing your accounts. This means that instead of budgeting and detailing for example, account number 6801 - Outside Services, New - we budget and detail for base account 68. We ignore distribution between the departments since how we allocate expenses to each department has nothing to do with profit or reducing expenses.

✓ Quick Tip - Spend less time figure out how to distribute expenses between the departments (who should pay of what?) and more time trying to reduce the expense.

Depending on your computer system, detailing your expense account usually involves an inquiry of the account. If possible get all the accounts at once and then cut and paste the detail into Excel. For example if you have R+R, you'd use job 0504 (use next to move to the next account), 0367, or 0368. For ADP, you can use AGRI. If you have some of the newer programs like ERA XT reporting or ADP's w.e.b.Suite series – or one of the new PC-based DMS systems– detailing accounts is much easier. You can normally cut and paste from the screen into Excel. There are also add-on products like ADP's AGRE with PBX to get a report of base accounts. In addition, you can write a query or RPG from your detail files; GL.JE.DTL for ADP with the account type of "E" or the file JOURNAL for R+R. For other systems use their reporting tool from the GL detail file and either use a range of accounts to get all the accounts in your base account, or pull all with a type of expense into Excel and you can filter in that product. A later section has information for ADP, R+R and other system users.

The following is an example of the detail for data services for all departments. We bring this back into base accounting because although $1500 a month for the parts manager doesn't seem like a lot – over $6,000 for the dealership is huge.

Acct Base	78

Sum of Amount		
Vendor Name	Description	Total
AMERICAN NISSAN	(blank)	$1,085.10
COMPUTER	COMPUTER	-$211.00
KELLEY BLUE BOOK	SOFTWARE	$568.00
REYNOLDS & REYNOLDS	(blank)	$3,274.49
SIGN FEE	REYNA SEP/DEC	$1,821.20
(blank)	COMPUTER	$300.00
Grand Total		$6,837.79

How to Create an Expense Pivot Table

Although detailing an expense account is great for looking for mistakes, we need to pull the data back into a concise format like the above example so we can figure out what is the account used for.

After downloading your data from your DMS system, create a pivot table of the data using the steps on the next page.

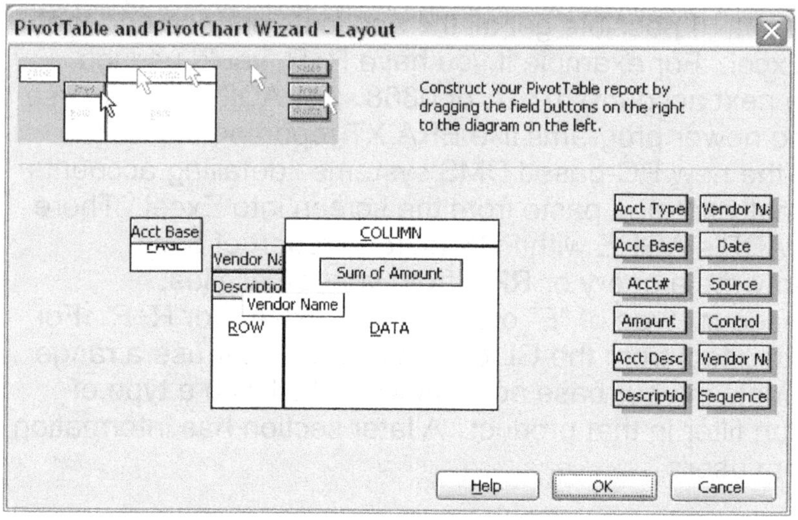

Select the whole worksheet (make sure each column of information has a header or title row.)

For Excel 97-2003 - From Excel's pull down menu, select **Data, Pivot table, Next, Next, Layout.**

For Excel 2007, Select a cell in a range of cells, or put the insertion point inside of a Microsoft Office Excel table. Make sure that the range of cells has column headings. To create a PivotTable report, on the Insert tab, in the Tables group, click PivotTable, and then click PivotTable.

Drag the amount into the Data – and select SUM. Drag the vendor name, posting description into the row. Put your account base into the Page. Click on Ok and then Finish.

At the top, select only one account number next to the Acct Base field (click on the pull down arrow in column B.

	A	B	C	D	E
1	Acct Base	78			
2					
3	Sum of Amount				
4	Vendor Name	Description	Total		
5	AMERICAN NISSAN	(blank)	$1,085.10		
6	COMPUTER	COMPUTER	-$211.00		
7	KELLEY BLUE BOOK	SOFTWARE	$568.00		
8	REYNOLDS & REYNOLDS	(blank)	$3,274.49		
9	SIGN FEE	REYNA SEP/DEC	$1,821.20		
10	(blank)	COMPUTER	$300.00		
11	Grand Total		$6,837.79		
12					
13					
14					

Pivot Table Format Tips

Right click anywhere **inside** the pivot table and select the Wizard to go back and work on the layout and format.

Double click on the amount and then click on **Number.**

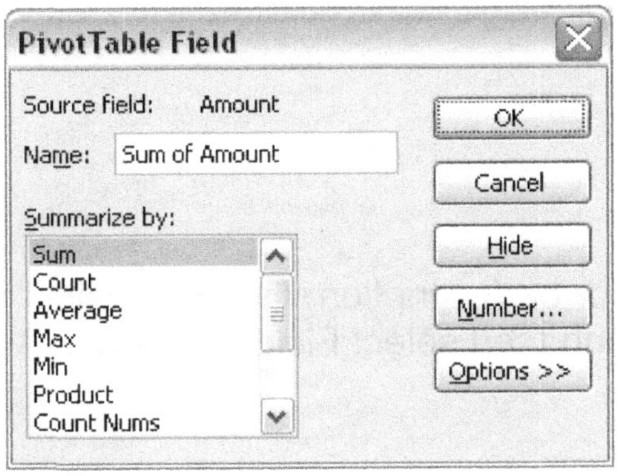

Change the format to Currency.

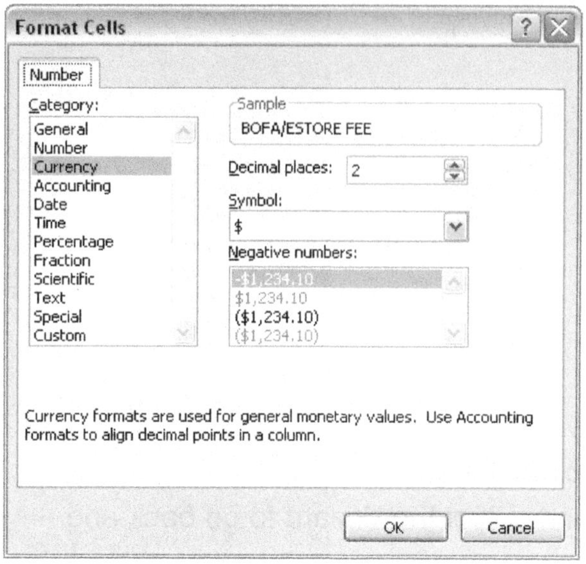

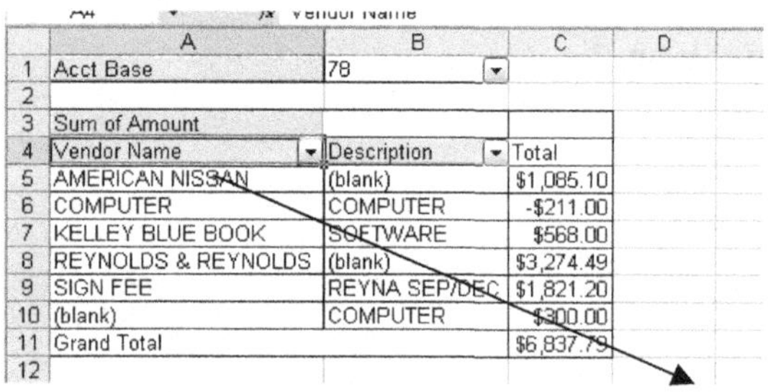

	A	B	C	D
1	Acct Base	78		
2				
3	Sum of Amount			
4	Vendor Name	Description	Total	
5	AMERICAN NISSAN	(blank)	$1,085.10	
6	COMPUTER	COMPUTER	-$211.00	
7	KELLEY BLUE BOOK	SOFTWARE	$568.00	
8	REYNOLDS & REYNOLDS	(blank)	$3,274.49	
9	SIGN FEE	REYNA SEP/DEC	$1,821.20	
10	(blank)	COMPUTER	$300.00	
11	Grand Total		$6,837.79	
12				

If you are getting subtotals (two lines for each vendor or description) then right click on any of the fields like Vendor name, description and then select Field Settings.

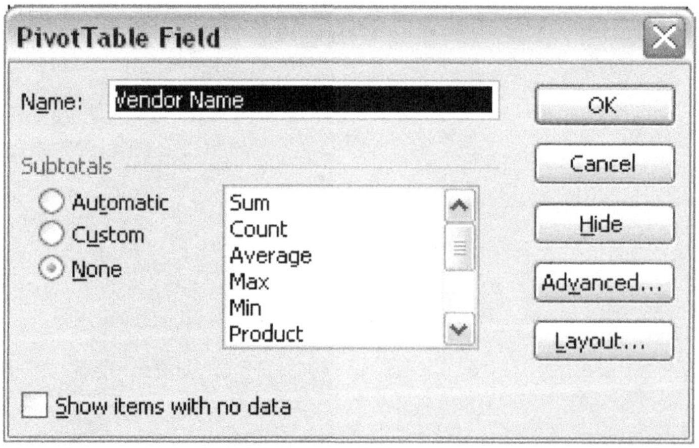

Click on Subtotals – None. It is important that your report is simple to read – if you can have the vendor name and posting description the report – that helps. As time goes by you'll work with your accounts payable clerk to post invoices correctly.

For example instead of this data;

Vendor: Reynolds + Reynolds
Posting Description: Inv 34344

You would have the accounts payable clerk put this in the description;

RR-Car Deal folders

Put a short name for the vendor and the **item** that was purchased. By using this focus on items that were purchased instead of vendors, you can focus on reducing these expenses.

Item Expense Method

Item expense analyzes is a new way to look at expense accounts. In the past we were trained to trend accounts to make sure everything looked right. That is great, but it only goes so far. Sometimes credits offset the actual increasing expenses – or a "big month" might cover up an actual expense that is creeping up on you. The key is the **Item Expense Method** to work on **only one item in one expense account at a time.** It is tempting to start attacking all your expense accounts, but this process only works if you work on one item at a time. Sometimes you can only work on one account each month. We'll also be treating each expense item separately so instead of trying to do hundreds of account numbers, you'll only do one – but EACH item charged to that account.

There are four ways to reduce expenses; the first and easiest is to eliminate the expense. That is not always an option; for example, employees need toilet paper. Second, you can change a process that will reduce the expense. An example of this is to pay off your floor plan earlier and reduce interest expense. The third way is tough. Many expenses increase because your staff takes it home or uses it for personnel use. The last way is to find a less expensive provider of the same expense. This requires creating an RFP, selecting a preferred provider, and retaining your employees to use only this vendor. Looking at the expenses that you detailed in the previous sections you can then divide your expenses into these four categories. Sometimes you'll need to do a combination of all three. For example with car wash soap, you might find a better way for the person to apply the soap to use less or prevent spills. Next you might lock up the supply to prevent it being taken home and last, you might find a lower priced source.

- ☐ **Eliminate – do we really need this?**
- ☐ **Reduce – is there a way to use less?**
- ☐ **Protect – is it being stolen?**
- ☐ **Bid – could we get it for a lower price elsewhere?**

Item Expense Worksheet

**** Click on Contact-Free stuff form at SandraJerome.com and click on the link for Super Controller zip file of spreadsheets.***

Expense Account Name:	Other Supplies
Account number (s)	61
Item description	Car Wash Soap
Vendor(s)	ABC Supply
Average monthly price$	$1297
Metric	$.50 per wash
Physical observation:	Two 5 gallon drums stored outside by the car wash
Eliminate – do we really need this?	Yes
Reduce – is there a way to use less?	1. Yes, it should be diluted (according the car wash manual and supply instructions) 2. It is being spilled when poured into the car wash. Get a funnel
Protect – is it being stolen?	Yes, it should be stored in the parts department

Understanding the Different Types of Expenses

Before you can budget expenses and create metrics, you need to understand what type of expense account it is.

Type of Expense	Examples	Notes
Variable	Sales commissions, F&I commission, some delivery expense	Must go up and down if less vehicles or service is sold
Fixed	Rent, owner's salary, depreciation	Same amount each month all year
Fixed - Seasonal	Utilities, outside services	In very cold or warm climates the amount varies greatly
Budget – Variable	Floor plan, employee benefits, and taxes	Budget is based on an account balance or other expense
Budget – Fixed	Policy work, personnel, data processing, repairs	
Reclassify		Amounts charged to the wrong account number
Other		

Creating a Metric for Expense Accounts

Metrics enable us to determine if the amount is reasonable or not. For example, sales commission might be 18% of the gross profit. Floor plan might be 6.5% of your new vehicle inventory balance. Utilities might be an average of $8,000 a month – going up to $12,000 in the winter and down to $4,000 in the summer. Each expense account needs some type of check and balance – otherwise the most we're able to say is "it looks higher or lower this month." To be able to instead say that it matches our metric – or investigate when it varies from the metric provides the confirmation we need to feel comfortable with the amount. ***Fill out the Contact-Free stuff form at SandraJerome.com and ask for Super Controller zip file of spreadsheets to get the*** Super Expense Worksheet to help you get started in budgeting with metrics – using the item method.

Super Expense Worksheet

Description	Item	Type	Actual
Sales Commissions	Commissions on Deals	Variable	$63,472.97
	Spiffs	Budget	$ 6,932.47
	Minis	Budget	$13,472.09
	Minimum Wage	Budget	$ 9,344.21
	Salary	Budget	$ 3,500.00

The left side of the spreadsheet is for you to enter each item that you found and select the type of Expense (it is a pull down selection.) You would then enter your accounts in the actual column.

Action	Budget	AvsB	Metric	Notes/Distribution
Protect	$59,077.84	$ 4,395.13	19%	of Gross profit, some deals worked at wrong rate
Reduce	$ 5,000.00	$ 1,932.47	0%	Do not pay out after $5000 is reached
Reduce	$ 1,098.60	$12,373.49	3%	Wholesale vehicles that can't be sold at a profit
Eliminate	$ -	$ 9,344.21	0%	Better training and distribution of deals
Eliminate	$ -	$ 3,500.00	0%	Two salespeople were stil paid salary after going on commission

Then you would pick on the actions for this account and determine the new budget amount. The AvsB column is an "Actual vs. Budget" calculation. The Budget is based on the metric you select times the key accounts. In this example the 19% is based on total vehicle gross that you would enter in your key accounts tab.

Total Vehicle Gross	$310,936.00

Keep going down the expenses creating a budget for each. Don't use the easy method like your 20 group numbers, a percent of gross or last year's average (although the next section has some guidelines to get you close.) Every expense needs to have a budget based on real amounts. If you have three lot people in other salaries, then the budget should be exactly their wages. Any variance points to a problem like overtime that needs to be investigated. Yes, this is a hard chore but start with a few accounts and continue from their. Have a detailed policy of what goes into that account and how you budget the amount and who needs to be alerted when you have a variance. Remember it only takes a tiny pebble to come lose in an earthen dam to cause the flood. Plug those holes before they happen and you'll retain more of your profit this year.

Some Common Metrics

*** Click on Contact-Free stuff form at SandraJerome.com and click on the link for Super Controller zip file of spreadsheets.*** The Expense Metrics spreadsheet that you can use to look at some various metrics. These vary by manufacturer, location and dealership size. For example a very small dealership with low gross profit has a challenge staying within personnel guidelines since you still need someone to answer the phone, cashier, and help customers. The utilities are much greater in Minnesota than in Oregon. Honda dealerships don't get floor plan credits but many GM dealerships make money on floor plan.

Expense Metrics		**Actual Gross**	$	397,093.27	
Description	*Guide*		**Your Amount**		Variance
Sales Commissions	**8.00%**	$ 31,767.46	$	47,321.04	$ 15,553.58
Delivery Expense	**0.50%**	$ 1,985.47	$	2,854.00	$ 868.53
Policy Work-Vehicles	**0.50%**	$ 1,985.47	$	2,750.00	$ 764.53
Total Selling Expense	**9.00%**	**$ 35,738.39**	**$**	**52,925.04**	**$ 17,186.65**
Salaries-Owners / Executive Managers	**2.00%**	$ 7,941.87	$	11,500.00	$ 3,558.13
Salaries-Supervision	**10.00%**	$ 39,709.33	$	67,323.00	$ 27,613.67
Salaries-Clerical	**3.00%**	$ 11,912.80	$	12,533.04	$ 620.24
Other Salaries And Wages	**6.00%**	$ 23,825.60	$	32,534.00	$ 8,708.40
Absentee Compensation	**2.00%**	$ 7,941.87	$	5,753.98	$ (2,187.89)
Incentives-Supervision	**1.00%**	$ 3,970.93	$	5,500.00	$ 1,529.07
Taxes-Payroll	**4.50%**	$ 17,869.20	$	28,343.03	$ 10,473.83
Employee Benefits	**2.00%**	$ 7,941.87	$	11,433.33	$ 3,491.46
Retirement Benefits	**2.00%**	$ 7,941.87	$	12,932.33	$ 4,990.46
Total Personnel	**32.50%**	**$ 129,055.31**	**$**	**187,852.71**	**$ 58,797.40**
Company Vehicle Expense	**1.00%**	$ 3,970.93	$	343.34	$ (3,627.59)
Office Supplies And Expenses	**0.25%**	$ 992.73	$	873.43	$ (119.30)
Other Supplies	**0.50%**	$ 1,985.47	$	4,734.33	$ 2,748.86
E-Commerence Advertising/Fees	**1.00%**	$ 3,970.93	$	6,934.33	$ 2,963.40
Advertising	**4.00%**	$ 15,883.73	$	39,534.33	$ 23,650.60
Advertising Reimbursements	**-1.00%**	$ (3,970.93)	$	(233.44)	$ 3,737.49
Contributions	**0.01%**	$ 39.71	$	50.00	$ 10.29
Policy Work-Parts And Service	**0.50%**	$ 1,985.47	$	3,745.33	$ 1,759.86
Information Technology Services	**2.00%**	$ 7,941.87	$	11,543.00	$ 3,601.13
Outside Services (Other)	**2.00%**	$ 7,941.87	$	14,344.33	$ 6,402.46
Travel And Entertainment	**0.25%**	$ 992.73	$	5,434.33	$ 4,441.60
Membership Dues And Publications	**0.10%**	$ 397.09	$	975.00	$ 577.91
Legal And Auditing Expense	**1.00%**	$ 3,970.93	$	6,533.00	$ 2,562.07
Telephone	**0.50%**	$ 1,985.47	$	975.33	$ (1,010.14)
Training Expense	**1.00%**	$ 3,970.93	$	254.00	$ (3,716.93)
Interest-Floorplan	**1.00%**	$ 3,970.93	$	11,934.00	$ 7,963.07
Interest-Floorplan Credit	**-2.00%**	$ (7,941.87)	$	(10,304.00)	$ (2,362.13)
Interest-Notes Payable (Other)	**1.00%**	$ 3,970.93	$	8,744.31	$ 4,773.38
Insurance - Inventory	**0.50%**	$ 1,985.47	$	1,147.32	$ (838.15)
Bad Debt Expense	**0.10%**	$ 397.09	$	657.44	$ 260.35
Freight	**0.50%**	$ 1,985.47	$	(357.00)	$ (2,342.47)
Miscellaneous Expense	**0.10%**	$ 397.09	$	788.01	$ 390.92
Total Semi-Fixed	**14.31%**	**$ 56,824.05**	**$**	**108,650.72**	**$ 51,826.67**

Rent	**5.00%**	$ 19,854.66	$ 18,500.00	$ (1,354.66)
Amortization-Leaseholds	**0.25%**	$ 992.73	$ 1,500.00	$ 507.27
Repairs-Real Estate	**0.25%**	$ 992.73	$ 78.43	$ (914.30)
Depreciation Bldgs. & Improvements	**0.50%**	$ 1,985.47	$ 3,743.44	$ 1,757.97
Taxes-Real Estate	**0.50%**	$ 1,985.47	$ 9,374.33	$ 7,388.86
Insurance Bldgs. & Improvements	**0.25%**	$ 992.73	$ 1,187.05	$ 194.32
Interest - Mortgages	**0.10%**	$ 397.09	$ -	$ (397.09)
Utilities	**2.00%**	$ 7,941.87	$ 7,143.44	$ (798.43)
Insurance Other	**0.25%**	$ 992.73	$ 5,732.09	$ 4,739.36
Taxes-Other	**0.25%**	$ 992.73	$ 594.04	$ (398.69)
Repairs-Equipment	**0.01%**	$ 39.71	$ 89.43	$ 49.72
Depreciation-Equipment	**1.00%**	$ 3,970.93	$ 1,593.04	$ (2,377.89)
Equipment Rental	**0.01%**	$ 39.71	$ 932.04	$ 892.33
Total Fixed	**10.37%**	**$ 41,178.57**	**$ 50,467.33**	**$ 9,288.76**
Total Expenses	**66.18%**	$262,796.33	$ 399,895.80	$ 137,099.47
Net Profit		$134,296.94	$ (2,802.53)	

In this spreadsheet example the dealership makes almost $400,000 in gross profit but loses a couple thousand each month. If they brought their expenses into an industry guide – they'd make $134,296 without increasing gross. That is just how important expense control can be. Open in Excel the Expense Metrics file. Enter your gross profit and your actual expenses in the "Your amount" column. Adjust the metrics to your guidelines and then use this monthly to find target accounts to apply the item expense method to.

A Tutorial of Item Expense Analysis – Variable account

Let's now put everything that we've learned so far together into a tutorial.
Sales commissions should be budgeted as a percent of gross profit after pack.
In this example if the dealership brings their sales commissions in line with industry metrics, it will have over $50,000 more profit per month – or over half a million a year.

Step 1 – What is your metric for this expense account? I'm going to use 23% of gross profit as a starting point.

Pay Plan Analysis - Step 2	
Gross Profit	$ 1,942.00
Pack	$ 450.00
Payable Gross	$ 1,492.00
Percentage	25%
Commission	$ 373.00
Percentage	**19%**

Step 2– How does your pay plan really pay? In my example below, the Pay Plan on an average deal should net at 19% based on the average gross profit less pack and coming up with a payable gross of $1492. The commission is $373 which is 19% of the dealership gross.

Sales Commission - Step 1	Benchmark		**23%**	
Step 3 - Car Deals		**New**	**Used**	
Gross Profit	$	232,417.00	$297,342.00	
From car deals	$	63,433.00	$ 87,343.00	
Percentage		**27%**	**29%**	
Deals paid per pay plan				
Calc commissions gross	$	192,323.00	$201,343.00	
Commissions	$	39,332.00	$ 37,343.00	
Percentage		**20%**	**19%**	Okay
Mini Deals	$	40,343.00	$ 95,999.00	
Commission	$	27,333.00	$ 43,232.00	
Step 5 - Apply Benchmark		**68%**	**45%**	
Total Commissions	$	79,191.00	$ 95,980.00	
Percentage		34%	32%	
Benchmark		23%	23%	
Profit Aim	$	25,735.09	$ 27,591.34	
Total Profit Available	**$**	**53,326.43**		
		12		
	$	639,917.16		

Pay Plan Analysis - Step 2	
Gross Profit	$ 1,942.00
Pack	$ 450.00
Payable Gross	$ 1,492.00
Percentage	25%
Commission	$ 373.00
Percentage	**19%**

Step 4

Other Expenses from payroll		
Spiffs	$ 1,850.00	$ 2,950.00
Bonuses	$ 3,433.00	$ 11,223.00
Minimum wages	$ 7,243.00	$ 1,232.00
Total Other	$ 12,526.00	$ 15,405.00
Percentage	**5%**	**5%**

Step 3 – What is the expense from car deals? In my example it is 27% and 29%. Way off of our pay plan and metric. That means you'll need to break out your car deals. This is a tough step. First download your new and used journals or query your journal files into Excel and isolate the gross profit and commission for every deal. Next break them out by calculated commissions (those that were paid per the pay plan) and minis (those that were paid a flat amount due to a low gross profit or loss.) As you can see in my example the calculated commission came out perfectly with the agreed upon sales pay plan – 19%. The problem came from the minis. One problem is now found. You must accrue sales commission on the car deal posting for this step to work.

Step 4 – What are the other expenses charged to that account? Add up all the spiffs, bonuses, and minimum wages. In my example, this came to 5% which with the pay plan of 19%, we're one percent over metric. Look at the various amounts and see if you can budget to come in line. All of these are controllable – make a budget for spiffs each month and only allow a salesperson to be paid minimum wage for a set amount of months before providing more training, coaching – or terminating.

Step 5 – Apply the metric to your gross profit and actual commission and find out how much profit you're missing the target with by not meeting the metric. In this example it is over $50,000. That is enough of a miss off the mark to make a plan for profit. Since your biggest problem is mini deals, you need to analyze each of those deals. Pull them out. How did it happen that you couldn't make a decent profit? For new vehicles – what happened? Who ordered them? Did they come from dealer trades? Did you add too much equipment or keep them as a demo for too long?

For used vehicle – should these have wholesaled instead of paying that sales commission – especially if they weren't vehicle in your product line that could convert to service customers? Who appraised the vehicle – or worse – who purchased it? Time to find out what needs to be changed – remember, over half a million is at stake! **** Click on Contact-Free stuff form at SandraJerome.com and click on the link for Super Controller zip file of spreadsheets.***

Example of a Budgeted Expense Account

Advertising is a perfect example of a budgeted account. It must have a budget and purchase orders created. It is the 2nd most important item to perform an actual vs. budget after sales commissions. If your sales managers are having difficulty sticking to their budget, move the purchase order function to the controller and do not allow advertising without a purchase order. Advertising is not a variable expense. At the end of the month it does not go down like commissions if you sold fewer units. Consider it a fixed expense and budget a fixed amount. If you have to be the one to control the purchase orders, then do it. This form is in Free downloads at SandraJerome.com and you can use it for any budget account. The 2nd tab has an analysis of Budget vs. Actual and enables you to enter your advertising sources to see if you're spending the money where you get the results.

ADVERTISING BUDGET vs. Actual							
Complete after all bills are entered							
Month:							
					UP	% of	Effectines
					Source	Total	Variance
TV	Budget	Actual	%	Variance			
KONP	$ 17,000.00	$ 23,427.00	47%	$ 6,427.00	16	18%	-30%
KOMO	$ 9,000.00	$ 9,247.00	19%	$ 247.00	27	30%	11%
$ -	$ -	$ -	0%	$ -		0%	0%
$ -	$ -	$ -	0%	$ -		0%	0%

ADVERTISING BUDGET/Purhase Order

Please Turn into Office Manager/Controller by the 1st of the Month

Month: Aug-08 **Reference** ________

Source____
Posted ___

TV	A/P #300 Ctl#	Credit	Debit Acct			
				Total:	$ 26,000.00	100%
KONP	11KONP	$ 17,000.00	11A	New	$ 7,800.00	30%
KOMO	11KOMO	$ 9,000.00	11B	Used	$ 7,800.00	30%
			65D	Service	$ 7,800.00	30%
			65F	Parts	$ 2,600.00	10%
				Other	$ -	0%

Radio	Ctl#					
KSPW	11KSPW	$ 2,300.00		Total:	$ 11,500.00	100%
KWRP	11KWRP	$ 9,200.00	11A	New	$ 3,450.00	30%
			11B	Used	$ 3,450.00	30%
			65D	Service	$ 3,450.00	30%
			65F	Parts	$ 1,150.00	10%
				Other	$ -	0%

Print	Ctl#					
Star News	11STAR	$ 4,323.00				
Auto Trader	11AUTOT	$ 1,123.00		Total:	$ 7,846.00	100%
Toss Away	11TOSS	$ 1,500.00	11A	New	$ 2,353.80	30%
Welcomebook	65WEL	$ 900.00	11B	Used	$ 2,353.80	30%
			65D	Service	$ 2,353.80	30%
			65F	Parts	$ 784.60	10%
				Other	$ -	0%

Other	Ctl#					
				Total:	$ 421.00	100%
US DEX	65DEX	$ 421.00	11A	New	$ 126.30	30%
			11B	Used	$ 126.30	30%
			65D	Service	$ 126.30	30%
			65F	Parts	$ 42.10	10%
				Other	$ -	0%
				Total	$ 45,767.00	

Manager's Signature: ____________________ Grand Total $ 45,767.00

Expense Accounts – Over 100 Reduction Tips

Do you have a suggestion box to enable employees to propose expense reduction ideas? Consider having a monthly best idea award and print the winner's idea in a newsletter that you attach to each pay stub.

Commission Expense

- ❐ Get rid of your commission vouchers. A complicated commission voucher system of paying commissions hurts your profits by enabling someone to overpay a salesperson either by stealing or by an honest mistake. The voucher forms cost money, the commission summary sheets cost more money and both are labor intensive and are full of mistakes (we always catch the underpayments, but the overpayments go by undisclosed.). If you pay commissions at 25%, then why not run a report of the car deals for last month, provide the totals by salesperson and then pay them 25% of the total?
- ❐ Find out why you have chargebacks and stop them from happening. If repair orders are left open too long, then close them quicker (why are you waiting for the hard copy?) If charges are not getting added to vehicles – use an automated PO system and due bills. Chargebacks cost you more than just the overpayment of commissions – they have a negative effect on the sales force and that can hurt sales. There is no excuse for chargebacks – your sales managers should not be selling vehicles that we do not know the costs for. This is a much larger problem than trying to make sure we recover the commission expense.
- ❐ Reduce Mini deals. If you are forced to sell at a loss then wholesale or "fleetail" the vehicle instead of retailing. If it is a new vehicle then try to use one of my campaigns in the Ultimate Campaign Guide to get rid of those vehicles to vendors, employees and family. Other old new vehicles could be given to your BDC or Internet department to sell at cost. You should not retail at a loss and mini deals encourage this. Think about eliminating them from your pay plans. If you start paying 25% of the salesperson's total gross generated for the month, then the fact that some deals had a smaller gross will not be as noticeable in a report as they appear on an individual commission voucher. Mini deals throw your variable expense percent off.
- ❐ Accounting gross must equal sales gross. You need to run a report from your F&I system of the payable gross profit for the month. You say that the F&I gross is wrong? This can't be! You must get accounting gross to match F&I gross then create the fields to correctly calculate payable gross. Your first step is to see if it's a problem by running a report that compares accounting gross to F&I gross. I've actually been at a dealership where these two amounts were exactly the same. I was shocked and when I asked the

controller how he accomplished this – he merely responded, "Why shouldn't they be the same?" Perfect answer! His system was that when the billers in the accounting office posted a deal they made sure that F&I had the same gross. If they didn't match, she stopped and questioned the sales manager or F&I manager and made sure sales and accounting were in sync on the deal's gross profit. Yes, we can say that "the sale department is never right," but that is scary. What if accounting missed a rebate and even missed submitting for a rebate? That profit is gone forever. See how commissions can affect profit? If this comparison of F&I gross to accounting gross is done weekly, then you can run it one more time after the month is closed and then run your commission report and pay commissions. That's another issue – I like doing sales commission on the 10th or 15th of the next month. During the month a salesperson gets a draw equal to their minimum wage or salary, we close the month with accrued commissions from the deal postings and settle out the commissions between the 10th and 15th. All your held deals should be cleared by then and any adjustments to gross should be done by this time.

Delivery Expense

- ❒ Close internals every day. If the repairs go to cost of sales instead of policy, the managers will "sell" the repairs and you'll deduct from commissions.
- ❒ What supplies are being used to clean vehicles? Can you improve how you use the supplies (use less?) Can you purchase for less? Are they being controlled by the parts department (avoid being taken home?)
- ❒ Do you do unnecessary safety inspections? Determine immediately if a vehicle is being wholesaled and avoid this expense along with cleaning, gas, and detail.
- ❒ Why do you remove equipment from vehicles instead of having your parts department order a new option and have the service department install? You'll make more money since it is hard to recover the cost of removal.
- ❒ Determine exactly what needs to be done to a vehicle to prepare it for delivery and make sure the expense matches the units you delivered.
- ❒ Are you filling the tanks of "their dealer trades?"
- ❒ Make a policy for how much goes into each type of delivery and vehicle for the lot. For each "fill the tank" there must be a new delivery.
- ❒ Where are the credits going? Reconcile that to the vehicles purchased each month.
- ❒ What is the cost of missing equipment each month? Do you have a key control system to know who was in that vehicle that month? Just by letting employees know that you tracked this helps reduce losses.
- ❒ Are salespeople losing the security fobs for key systems? Make them give you their keys before they can check out a security fob and they'll be sure and return it to get their keys back.

- ❒ Can you reduce your dealer trades? For one month analyze all dealer trades and find out how many similar vehicles that you had in stock and what were the reasons for the trade (wrong color, wrong equipment, etc.) Could you order vehicles better?

Policy Work – Vehicles

- ❒ Analyze your policy work; how many days since the sale did it happen? Does one salesperson have more policy on his deals than others? Many times policy was "promised" in the deal and should have been a due bill and a cost of sale.
- ❒ Who is purchasing these vehicles that need so much policy? Are they being bought online or by a buyer who does not inspect or test drive the vehicle?
- ❒ Who authorized policy work? If you pay your advisors on policy work and let them authorize it – then that is giving them a blank check. Require a purchase order from the sales department (for new and used) and the service manager for service policy.

Demonstration

- ❒ Use your showroom control or CRM software to tell you how many demos you had that month and calculate how much each demo ride gas should be.
- ❒ What is the route they take? Could it be shorter?
- ❒ Can you find a supplier for gas that charges less or uses a control system that enables you to enter the stock# and user for each gas purchase? Consider getting separate cards for each department to accurately watch this account.

Advertising

- ❒ Create a monthly budget for advertising (as in the previous section) and cut off the account when the amount has been spent. This requires using a strict purchase order system and the controller might have to be the one to issue purchase orders to your advertising suppliers.
- ❒ Yellow pages are becoming obsolete. Consider just getting the one line.
- ❒ Are you using all your co-op funds? Who is responsible for submitting?
- ❒ What about your association advertising fund? Are there unused credits available for you?

Interest - Floor Plan – Vehicles

- ❒ Create a report of date sold vs. date paid off for your new and used floor plan. Find the delays and payoff units faster
- ❒ Have the F&I managers submit the contracts to save interest expense.
- ❒ How many days does it take from received date to have a vehicle on the front line ready to be sold? Reduce this time to turn your inventory quicker.

- ❒ Did you have any vehicles delayed in shipment? Many manufacturers reimburse you for this excess time.
- ❒ All frozen cash costs you more floor plan interest – so look for warranty claims not paid, AR past due, rebates that haven't been submitted.

Salaries – Supervision

- ❒ Have you reduced clerical and other staff so much that you have to carry more high priced managers than other dealerships have?
- ❒ Small and medium dealerships may not need the many "layers" of managers that larger dealerships have like F&I directors, fixed operations managers, and general sales managers. Add up your managers and compare to how many people they supervise. Does one employee have 3 managers/
- ❒ Know your State's overtime law and avoid paying overtime to managers who are paid salary.

Salaries – Salespeople

- ❒ If a person is truly a salesperson, then their pay is commissions, not salary.
- ❒ What is your training program? Ask any salesperson that gets paid minimum wage instead of commissions about their training; what has been completed and who is in charge of training them.

Salaries – Clerical

- ❒ Clerical people are not as effective after 8 hours of detail work. Try to eliminate all office overtime except for cashiers that require "coverage." You might want to turn those late hour pick ups over to the advisors instead.

Other Salaries & Wages

- ❒ Know your State's overtime law and avoid paying overtime just because someone is clocking in early.

Absentee Wages - Productive Personnel

- ❒ Do you track vacation time? In the Free downloads at SandraJerome.com is a vacation accrual spreadsheet to help you keep track of this.

Taxes – Payroll

- ☐ Consider using a common Payroll paymaster – and use the corporate or one of the dealership accounts to pay all payrolls. This will save you the unemployment and FICA for employees that switch from one company to another and solve many of your check printing issues. Be sure to discuss this with your CPA firm.

- ☐ Make sure you have employees classified correctly for worker's comp and can you exclude your owner and his family? Most corporate officers can also be excluded, especially if they are family.

Employee Benefits

- ☐ Are you collecting the employee's portion of the medical insurance correctly? When pay the bill, charge it directly to their AR account to make sure.
- ☐ Why do you need a catered hot lunch on Saturday? Having sandwich making supplies or pizzas has the same effect of keeping them in the store. A large catered lunch keeps the salespeople in the lunch room for more time than if they had run out and grabbed a quick sandwich.
- ☐ Most other employees don't need to take the bank, insurance company CPA firm and other vendors out to lunch. As a custom, the dealership is their client and the other party normally pays. Question any reimbursement for lunches.
- ☐ Do you really need to drug test that many people that often? If you drug testing is the same day each week or month, the employees have it figured out, so rotate the date and pick fewer. Random is still random even if you do 5 people instead of 20.
- ☐ Do you really need donuts? They are causing insurance rates to increase. Consider supplying your customer only with sealed snacks from Costco instead.
- ☐ Water should be provided but it does not need to be bottled water. Consider a filtered system that uses your own water supply.

Pension Fund / 401 K

- ☐ Watch for high fees for this service. Most investment companies want this business and will reduce the fees.

Company Vehicle

- ☐ Make a list monthly of which vehicles are company vehicles; if you have a plow truck – do you need it all year round?
- ☐ Have the parts driver keep a log for one week. Is he running errands for managers?
- ☐ Consider getting the bank to pick up your deposits. Many will do it for free.

Small Tools

- ☐ Ask a seasoned technician or shop foreman to approve the small tool purchases on your factory statement. Some of that stuff comes automatically unless you call and cancel it. I have found dealerships getting small tools for models they don't even service (like medium duty trucks.)

Freight & Express

- ❒ Are you shipping contracts overnight by default when often 2nd day gets there in the same amount of time (especially within the same state?)
- ❒ Get a pricing comparison between DHL, FedEX and UPS. They will normally bid the rate to get your business.
- ❒ Is freight getting charged out on warranty claims, repair orders and parts tickets? Find out the freight policy and help counter people be better about adding this amount. Consider a freight bonus.

Stationery & Office Supplies

- ❒ One of the biggest paper wasters is the printing of schedules for reconciliation and storage. Could you go paperless? First of all, let's look at the printing for storage. You send them your month end backup tape and they send you a CD. No need to print your GL, journals and especially those bulky schedules.
- ❒ What about printing your schedule for reconciliation? Couldn't you view the schedule online? First, download your schedules to Excel. Next, look for posting errors by sorting by the amount – looking for all the unusual credit on a receivable schedule and debits on payable schedules. Then use your Find feature to find the debit or credit that you need. Highlight the full line of each entry, copy to a new sheet and that is your Journal Voucher (remember to reverse the signs if necessary.) Continue making your entries on the sheet, add some notes, print it and post. No more need to order those expensive preprinted Journal Voucher forms either.
- ❒ Are you paying for DMV and bank forms that you could get for free?
- ❒ How many copies are you making of forms? With a simple call to your computer vendor you can get rid of the extra copy that goes in the waste.
- ❒ Order lighter paper. If you still using hard copies for your repair orders, the typical paper is 32#. This premium paper costs four times what normal 20# paper does, and even more if you're getting the hard copies printed a buff color with lines and spaces for notes and flag sheets. With electronic dispatch and technician terminals, the need for these heavy hard copies has been decreased and most can be eliminated – or at least downgraded to a nice inexpensive 18-20# paper. For some of you, the 20# paper is free due to your laser fee contract, but you're getting charged a ton for the premium paper. Either way, you'll also save on shipping costs.
- ❒ Efficiency experts estimate that each check that you type costs between $2.00 and $8.00. This includes the cost of the paper, printing costs, ink, bank fees, the labor to produce, file, and reconcile the check to the bank. In addition, there are the postage and envelope to mail and the eventual storage costs. For any dealer or controller that has a sore hand from signing them – checks are a horrible waste. In addition, the data on a check can be stripped off with a fluid, leaving you open to theft. Can you reduce the amount of checks that you type? Most of Europe is on electronic payments and creating

checks in a business environment is becoming rare. Here are some ideas for you to check out;

- ❒ Refund checks – if your state requires that all refunds go to the customer, no matter how small, why not create a form that enables the customer to donate the refund to a selection of charities if the amount is less than five dollars? You can even match that amount, if you want to get a little goodwill out if it. Instead of typing thirty refund checks a month to customers, you could type one or two a quarter to a charity. Check with your legal and tax advisors for your state, but this might save you a lot of money and give the customers a warm feeling about themselves and you. Do they want to know how much was donated in their name? Easy, get their email address and you'll email them the amount. Another great way to collect email addresses. You also need to find out why so many $1.00 checks? If you F&I managers are tossing in a dollar or two to the license fees just to make sure it won't come out short, you need to stop it. It is costing you more than you're saving.
- ❒ Weekly vendors – if you have a vendor that wants to be paid weekly, stop that. Not only does it take extra time for your office to reconcile four statements a month, the interruption and cost of the extra 3 or 4 checks is too much. Is this happening? Run a report of all the checks created last year. All vendors should have a maximum of twelve.
- ❒ Payroll Direct Deposit – yes, I know most DMS systems do not have this option, but you don't need it. Find out the format your bank wants the check data in and write a custom report to upload the payroll information to your bank.
- ❒ Pay by credit card – if you have certain vendors that you trust, like the cell phone and trash company, pay that by automatic credit card. Add a bank account in your DMS for the credit card transactions. The debit would be to the normal expense account and the credit to accounts payable. Then run an A/P batch using this credit card bank account instead of your general account and print on blank paper if you must. When the bill comes in from the credit card company, type one check and debit the credit card bank account. This is the way Quickbooks works; they treat credit card accounts like bank accounts. You can also make the credit card company an A/P vendor – but make sure you put the actual payee in the description or control2 field so you'll get a good detail of your expenses. Don't forget to use a card that earns you flight miles or bonus points. If the credit card automatically debits your bank account monthly, then you'll need to review that statement for internal control.
- ❒ Why maintain the inventory jackets? Everything in the jacket is in the computer. The jackets are expensive to order and expensive in labor to fill in the VIN#, year, make, etc. Now that we keep the MSO and title in a fire proof safe in a binder, (I hope!) you don't need to maintain these jackets. You are especially causing a lot of extra expense by making copies of the repair

orders, bills of lading, and invoices. Think of all the expense to then enter all the charges to the vehicle and keep a running total – the same thing you'd see if you looked up the vehicle on the computer.

- ❒ Manual purchase order books are a thing of the past. Most DMS systems have automated purchase orders as part of their core software. Teach employees how to use this and you'll prevent chargebacks on vehicle sales, sublets from not getting charged out, and reduce posting errors. You'll eliminate an expensive office supply and increase productivity. You'll probably already paying for the software – so why not use it to save money?

Outside Services

- ❒ Are you being charged for any security systems that are no longer being used? Get a report from the company of the logged in and out times for each system to make sure someone is still setting them.
- ❒ Consider switching to one bank account for all dealerships. You still create the transactions and have 4 cash in bank schedules – but you only PRINT on one checking account and one check# in the printer. You use the 4 GL accounts to reconcile to the bank. This should fix the problem of transactions printing on the wrong checks. You can keep the deposits going into the 4 banks – this helps identify NSF checks and other charges.
- ❒ Do you really need a check guarantee service? If they take the check even when the guarantee service denies the check – why even have it?
- ❒ How many jump and vehicle pushers are you renting? Do you need that many and do you the ones that you're paying for?
- ❒ Compare your cleaning service contract to the billing. Are they really cleaning with the frequency being charged? Could you split up the work (like cleaning the floors) and sub contract to a less expensive company?

Telephone

- ❒ Consider IP/ Network phones that enable quick connections between dealerships – especially if you're in different toll zones.
- ❒ Are you using a current long distance code system? Has the codes been changed recently?
- ❒ Does the fax machine need a code? Make sure the handset is removed so employees can't use it as a direct dial phone.
- ❒ Take an inventory of all the phone lines and make sure each on is being used (and that you need it.) Many modems have been replaced by Internet access (or can be.) Check the computer room too.

Data Processing

- ❒ Take an inventory of all technology. Make sure you have logins for every software product and that somebody is using it to make money.

- PC support. Many dealers are paying close to $500 a month to support their PCs. You can buy a nice PC for $299 - so every month you could buy a backup PC and backup one of your older PCs to the new PC and keep it around in case the old PC dies. Be sure and get some easy backup software.
- Document management - at $700-$900 a month this is a lot to pay to store your documents. You can do the same thing with Kodata or CD-One view for $150 a month and if you leave your DMS later, you won't lose all that storage (unless you keep their system and support going, it's all gone.)
- SPG - Service pricing guides. Ask the advisors if they are using them. They cost between $200 - $600 a month and I find that most advisors don't like them or don't know how to use them. Don't let them tell you that they use them - ask to see how they work and make sure they can show you.
- Printers – most dealers pay a lot for printer support and like the PCs, could you buy a backup and support them yourself?
- If you are a R+R dealer and own Intellipath – make sure you've been trained on how to use. Look at your service repair orders from last year and yesterday. Is it the same coupon? Are you using the marketing features? This product can pay for itself – so sometimes instead of getting rid of a technology to save money – why not use it better to make more money?
- Normally a factory interface is required for any user that must DCS data from your DMS; warranty claims, delivery reporting, parts orders, and financial statements. Employees that do vehicle and parts locates, look at rebate programs, order vehicles – normally do not need an interface to the DMS system. Go to each employee and ask them how they access the factory and why. If they are going directly to a factory website login and never sending or receiving data from the factory to the DMS system – then they might not need that interface.
- Consider thin clients instead of PCs. A thin client is a network computer without a hard disk drive, which, in client/server applications, is designed to be especially small so that the bulk of the data processing occurs on the server. Thin clients cost less than a PC and they are easier to maintain. In a Thin Client environment you can upgrade users quickly and inexpensively by updating application software hosted on servers, not clients. Not only do you have a more controlled and secure environment, but technical support costs are greatly reduced. Thin clients are long-lasting products with no moving parts, to not only reduce noise levels, but also lower maintenance costs. There's even optional wireless connectivity support on select models. For about $350 you can get a thin client that lets you control software, applications and files centrally, save up to 80% on maintenance costs, and deploy out-the-box in under 10 minutes. In addition since they are technically "dumb PCs" you can reduce hardware theft by having something that nobody wants to steal – and if you do steal it – they don't walk away with your confidential files. Why not see if you can replace your next dead PC with a

thin client? Here are the numbers; 80% less maintenance per year, 25% capital cost savings, 34% less in maintenance, 23% less to operate, and 25% productivity increase (no games on the thin clients!)

- ❒ Consider OpenOffice instead of the pricey Office upgrade. OpenOffice.org is a multiplatform and multilingual office suite and an open-source project. Compatible with all other major office suites, the product is free to download, use, and distribute. What's in OpenOffice.org? For a Word Processor – you have **Writer -**. Features include; Auto-complete, auto-format, and real-time spelling check. For your Excel Spreadsheets, you can use **Calc**.
- ❒ Are you running your credit bureaus on your DMS system? Did you know that there are online credit providers like 700credit.com that charges about half the amount that you are currently paying? It has great features like the OFAC terrorist search and the ability to one-click to send a letter to your applicant. It's fast and your prospects are stored for access from anywhere you have an Internet connection. Since the software that you own from the DMS company can cost you $150 - $250 a month that's the first part of your savings. Next, let's say that you run about 800 credit inquires a month. If you save a dollar each, and eliminate the flat software fee – that's $1000 a month in savings – or $12,000 a year! In addition do you need to run 3 reports on each prospect? Isn't one report enough to figure out where to send the deal? Each report costs you money and many DMS system are set up to automatically run all three. Check you setups today for immediate savings.
- ❒ See if your bank offers remote check deposit – 11 banks currently do. All you do is run the check through a scanner and you get instant credit for the checks. The bank keeps a digital image so you can then destroy the original. It creates your deposit, and if you don't have any cash to deposit, you eliminate the need for an employee or a courier to take the deposit to the bank. Another potential benefit is that it cuts down on paperwork, and therefore reduces the chances of making mistakes or losing checks in the process of depositing them. Bounced checks also show up faster when processed through remote deposit.
- ❒ If you have multiple dealerships or locations you might be paying too much for "branch" accounts when you could operate under one login. First you need to determine if you're actually one company. I always use the tax return as my guide. Let's use as an example a dealership that has side-by-side stores – each with their own showroom. They have a combined service department and separate parts counters for each store. One is ABC Toyota and the other is ABC Kia. When I look at the tax return there is one return and it's a common ownership. When I look at their DMS billing I find two of everything – Service invoicing main and service invoicing branch. I go into the single accounting office and they are pulling two financial statements and constantly make inter company entries between the stores. The service advisors look first in one dealership and then log onto the other dealership. If you have one

tax return you should consider consolidating your DMS logins and use sub accounts to pay bonuses. In addition, if it's not a requirement of your manufacturer to show separate balance sheets (not sales reporting – that's easy) then consider combining them. You will save money and headaches.

Internet Expenses

- ❐ Reconcile every lead from your lead providers to the up log. Watch for duplicates from the same provider. You shouldn't have to pay for those.
- ❐ Are employees listening to Internet radio? That uses up your band width and cost you $$$.
- ❐ Do you really need a new T1 line for each manufacturer? Look at your Internet charges and see if they are necessary.
- ❐ How many websites are you paying for? Each one should provide you with web traffic reports to see if you need them.
- ❐ Are you paying two salespeople for the same Internet deal? The new breed of Internet salespeople can handle their own deals and no longer need a traditional salesperson to help them close the deal.
- ❐ Do you really need search engine costs? Having great keywords can reduce the need for these.

Repairs & Maintenance - Real Estate

- ❐ Consider using technicians for some of these services. Get a skill list from your techs and instead of paying for unapplied time, put them to work!

Heat, Light, Power & Water

- ❐ Find the water meter and watch it late at night or early in the morning. Leaks are common and you might have a small broken pipe somewhere.
- ❐ Do you really need all the lights on at night? Keep what you need for security, but start conserving energy. Your customers judge you by how green you can be.
- ❐ It is good for computers to be turned off at night. They clear out their memory and cache and don't use any power. It also helps against attacks.
- ❐ Turn off printer too – they often use more power than a PC!

Insurance - Other than Bldgs. & Improvements

- ❐ Get risk reduction ideas from your insurance company to help lower your rates
- ❐ Find out what type of claims (the amount and type) raise your rates. Consider self insuring for the small stuff and raising your deductable.

Expense Reporting Instructions for ADP Users

If you already own AGRE you may need to set up your base options; run PBX instead of PEX to get a base vendor analysis.

```
Would you like to:
   1.  Reprint the previous calculation
   2.  Recalculate Base Analysis ..............2

Which month do you want to print:
   1.  Current Month
   2.  Prior Month ............................1

Where do you want the report to print:
   1.  Line Printer
   2.  Terminal ...............................2

How would you like the report sorted:
   1.  By SOURCE by REFERENCE
   2.  By Descending Amount (with the highest dollars on top)
   3.  By VENDOR
   4.  By COMMENT .............................3

Enter Base Numbers to print or ALL for all Bases:
```

Hint – put in compresses mode first. You can also print a vendor analysis, but I find this report messy.

Detailing your Expenses RPG

Depending on how your staff enters data, you can normally get the information from this file, GL.JE.DTL. Ask them to be consistent about posting description and enter the item and vendor; for example – ABC Supply, car wash soap.

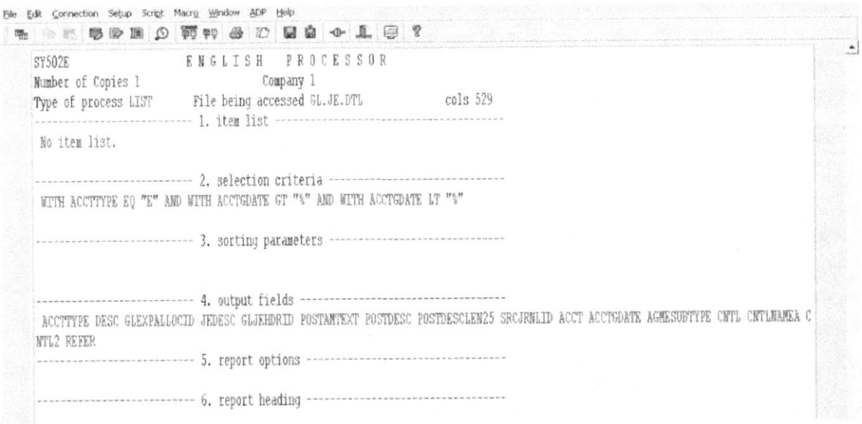

Expense Reporting Instructions for R+R Users

You might want to use 0504 to get the most detail, but you can also run a query to get the detail from the JOURNAL file. Make sure you run this report after all the expenses are in for the month but BEFORE you run month end.

File – Journal
With MMYY equal to the month you are working on
Output;
Account type
Account base
Account Number
Amount
Description
The posting Description
Month/Year
Source (journal)
Vendor/Control Name
Reference

```
ACCT.TYPE.DV   BASE   ACCT.NO   AMT   GL.DESC   DESC   MMYY  SRC   CTRL.NO
REF.NO
```

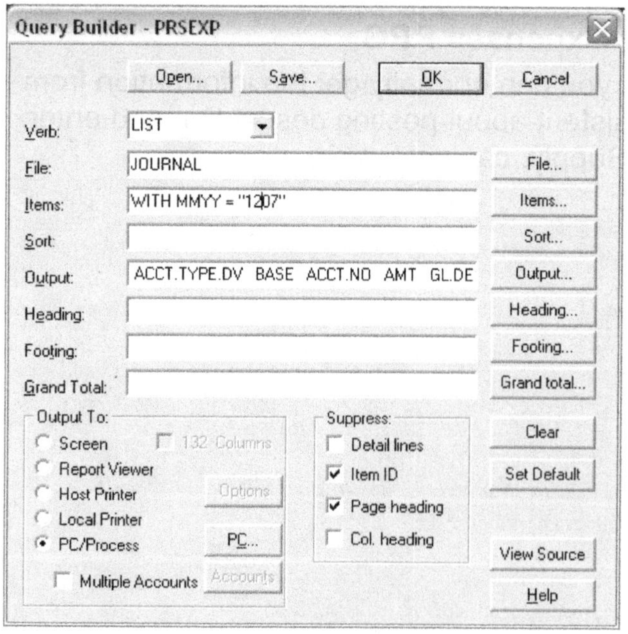

Expense Reporting – Other Systems

On most systems, when you see your detail for the account on the screen you can normally cut and paste the detail into Excel. Some of the systems allow you to "Copy Table" and then when you paste into Excel – it parses into each cell. If not, then when the data comes into column A, keep it highlighted and select in Excel, **Data, Text to Columns.** If you run a query of your GL detail file you want to get at least these fields;

1. Reference number
2. Date
3. Account number (6801)
4. Account number description (Outside services new)
5. Account type (try to only download the Expenses)
6. Account base (for some systems an account number like 6801 has a base account of 68.)
7. Control number and/or name
8. Posting description
9. Document description
10. Journal or source

If you link then this data to your master spreadsheet, you'll be able to rerun the report on your DMS system, download and then refresh the data in Excel without losing your formatting and formulas. Try to keep your formulas to the right and above the data.

Start with a new blank Excel worksheet. Click on the cell you want the data **Data, Import External Data, Import**

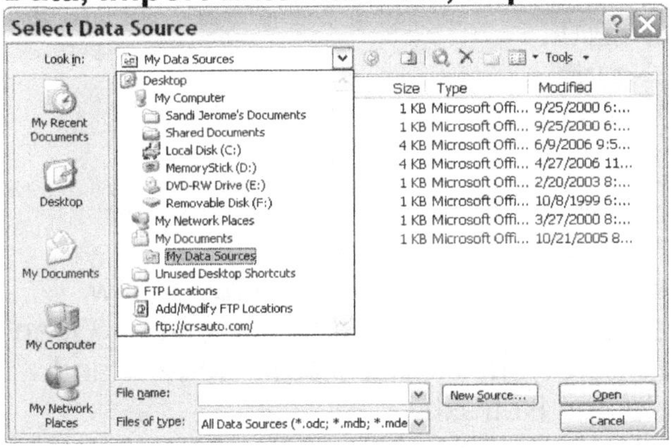

Change My Data Sources to the location of your download from your DMS system. Click on your file (that you created during your DMS system download.) Click OK

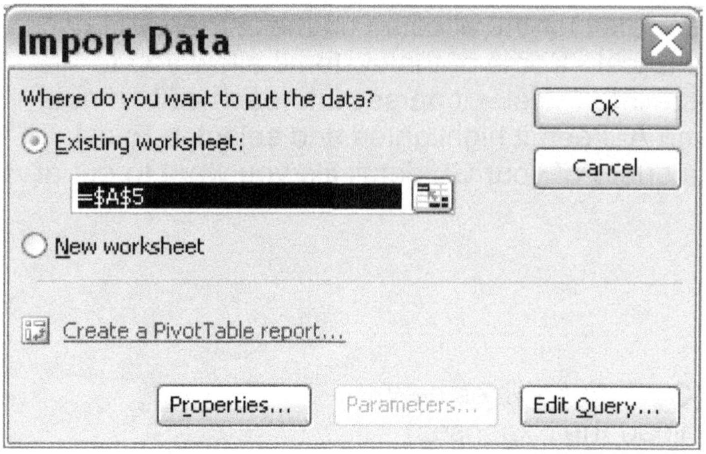

Bring to Existing Spreadsheet, A1

Click on whole spreadsheet – between A and 1 and Data, Pivot Table, Next, Next, Layout (to return to Layout later, right click anywhere inside the pivot table and select Wizard)

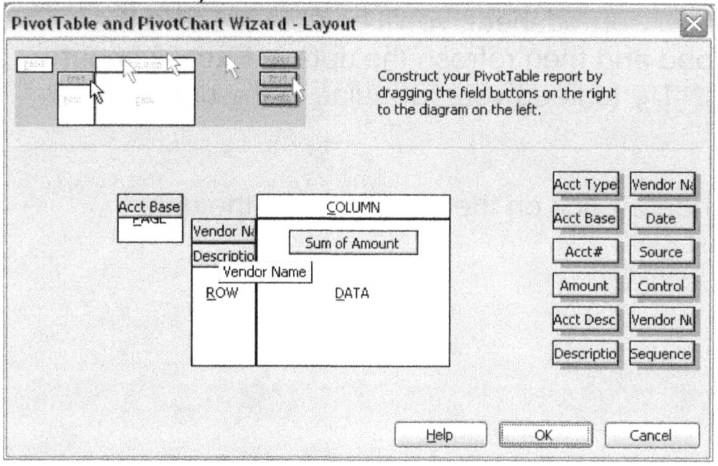

Now save this spreadsheet as your Expense Master – and when you run a new report from your DMS system, give it the same name as you did the first time. Open your Expense Master, right click on the tab where your previous data came in and select Refresh. Right Click on the pivot table and Refresh.

Part 2 – Employees

One of your largest expenses and productivity headaches are employees. This section of the guide is devoted to staffing, managing, and paying employees with the objective of getting results. I don't care how many employees you have if they are producing enough gross to justify keeping them. The key is to getting the most value from your personnel expense dollars.

Office Staffing and Job Descriptions

There are a lot of ratios out there for determining office staffing, but these are more of a guideline. How can you use these guidelines when no two dealerships are alike? One dealership might have the F&I office pulling contracts and another might dump that on the accounting office. Office staff might need to relieve the cashiers and receptionist – or that is done by service and sales staff. I prefer to create an office like you would a recipe, start with a core office of 5 and then start adding employees as you grow.

Small office structures – 0- 40 employees, 3-5 office staff

Office Manager w/accounts receivable
Accounts payable and payroll
New/Used biller/title clerk
Warranty/cashier
Receptionist/bookkeeper

After 100 deals, add another biller and split between new and used.
After 2 advisors, split cashier and warranty into two positions

Medium office structures, 41-100 employees, 6-7 office staff

Controller
Office Manager and Payroll
New Biller (100 retail deals)
Used Biller (100 retail deals)
Title Clerk
Accounts Payable
Accounts Receivable

At this point, the cashier and warranty would migrate to service employees working for the service manager and the office no longer relieves these positions. The receptionist works for the sales department and is relieved by sales personnel.

Larger office structures – Over 100 employees, 8+ office staff

Controller
Office Manager
Payroll and HR
1 - New Biller (per 150 deals/dealer trades)
2- Used Billers (per150 deals/wholesales)
1- Title Clerk
Accounts Payable
Accounts Receivable
Warranty
2 - Cashiers
Receptionists
* You can also combine Billers and Title Clerks into one position.

The above scenarios are merely guidelines for staffing your dealership based on various industry ratios. Keep in mind that no two dealerships are alike. For example one dealership might have an office person report deliveries and rebates to the manufacturer and in another it is done by the sales department. Some guides for an office staff is 1 office person per 15 employees not counting warranty, cashiers and receptionists. If you have less than 40 employees, then a five person office staff is normal. This includes the cashier, warranty, and receptionist – since the dealership is normally too small to have them separate from the office. If you are a larger dealership and do not count these positions as office staff, then a ratio of 16-17 to 1 is excellent. Even the smallest offices have an office manager position. Normally when the office grows past 7-10 employees, then there is an office manager and a controller. When the group adds multiple dealerships, then there is a group controller, or CFO. Most groups of 2 or more dealerships have a CFO. How does your dealership measure up? If you're having difficulty getting productivity from your office you might want to pay careful attention to my next issue where I'll give you some ideas about office bonuses. What haven't we taken into account? The salary you pay each person. Just like that unusual dealership with one office person, I know a biller that does over 300 deals a month – when the guide is only 100-150. She's highly qualified and could be a controller or office manager but enjoys being the best biller in the west! I would imagine that she makes more money than the biller that you have that was flipping burgers at Burger King in her last job. That brings up another method that I like to use to staff an office; a budget. Yes, I know that we're not very good at budgeting, but what if you made a budget for each position and then paid the office manager or controller a bonus for the amount under budget?

What does your Office Staff do All Day?

Nothing is more disappointing than finding out nobody was submitting rebates or reconciling one of your schedules. I find that having all employees complete a Job Function Worksheet enables you to see if there are any holes in your office (by consolidating into one big worksheet) and more importantly, realigning their functions to be related to each other. For example if you find 2-3 people handle car deals then move those functions into one person's tasks and move some of their functions to another person.

Job Function Worksheet

Your Name____________________

Current Position:____________________ How long?____ How long at this dealership?____How long in the car business?____ Do you have an accounting degree?___ Other degree________________

Dealership:________________________

Email address______________________________ Phone#____________

* Don't be concerned if the % of time doesn't add up exactly to 100% - just get close. ** Enter comments if you'd like. If you only do a portion of the function, cross out what doesn't apply

Function	**Percent of your time***	**I how to do this but not doing this now****	**I want to learn this****
Create daily bank deposit			
Verify daily cash sales and make sure it is zero			
Close daily receipts and match to bank deposit			
Take deposit to the bank			
Cashier ___ Service ___Parts			
Receptionist			
Answer Phones			
Send out AR statements			
Call past due AR accounts			
Research open AR items on the schedule			
Create daily open repair order report			
Add new AR accounts			
Add new AP accounts			
Heat sheet or report of AR past due accounts			
Research open warranty items			
Submit Extended warranty claims			
Receipt in down payments			
Receipt in payments for car deal			
Close repair orders – Cash and charges			
Close repair orders – Internals			

Submit warranty claims			
Post AP invoices ___parts ____vehicles ____sublet ____expenses			
Prepare AP checks			
Reconcile AP statements			
Sign checks			
Send contracts to the bank, GMAC, ChyrFin			
Post EFTs for contracts paid			
Post EFT for factory rebates and other credits			
Submit rebates			
Credit Life AH ___submit ___pay____refunds			
Figure gross profit and finalize deals on ADP			
Ext Warranty ___submit ___pay____refund			
Prepare sales commission vouchers			
Prepare sales commissions for payrolls			
Finance reserve statements, reconcile			
Clear inventory balances while finalizing deals			
Put open VR and CIT on a heat sheet or car deal due report			
Payoff/EFT floor plan			
Prepare jacket, key tags for ___New ___Used			
Payoff trades			
Create a jacket for trades			
Post vehicles into ADP ___New ___Used			
Stock in vehicles ___New ____Used			
Supervise cashiers			
Supervise phones			
Relief cashier			
Relief phones			
Supervise office staff			
Employee insurance			
OSHA claims			
Other insurance claims – accident reports			
Technician time sheets – total for payroll			
Payroll – hourly employee time cards			
Bonus payroll – accrue to pay plans			
Payroll calculate, update, print			
Payroll distribution between companies			
Payroll Qtr reports, submit taxes			
Sales tax report			
Factory Financial Statement – which one?			
Factory parts/monthly statement			
Floorplan reconcile			
Cash in bank reconcile			
Dealer trades – get paperwork ready			
Dealer trades – post			
Titling of vehicles ____New _____Used			
Wholesales deals – post			
Wholesales get paperwork ready for auction/wholesalers – receipt in payments			

Prepare other types of checks – type?			
Print factory communicates and disburse			
Open the mail and disburse			
Report deliveries to factory			
Issue purchase orders			
Send out mailers, flyers			
Make service appointments			
Order sales and service supplies (key tags)			
Order office supplies			
Repair PCs, printers			
Add users to the network, passwords			
Add users to the ADP system, passwords			
ADP system administration – uploads, tapes, backup			
Filing – what do you file?			
Other - list (use back if necessary)			

List all schedules that you "clean" each month ______________________________

List any GL accounts you reconcile monthly (prepaids, fixed assets, etc.)____

List any reports you prepare, for who and when (daily, weekly, monthly?)____

Please rate your level of Excel knowledge ___None ___Beginning ___Intermediate/Advanced

Please rate your level of ADP knowledge ___None ___Beginning ___Intermediate/Advanced

List 4 things that would make your job easier

1. ____
2. ____
3. ____
4. ____

Any other comments: (use back if necessary) – Attach your resume if available. (this form is in the free downloads.)

Creating a Training Program using Super Controller 3

After you have collected all these job function, it's time to create a training program based on these functions using Super Controller 3 Task spreadsheet– that lists over 300 tasks and tips on how to do. *** Click on Contact-Free stuff form at***

SandraJerome.com and click on the link for Super Controller zip file of spreadsheets. Tasks is one of the spreadsheets.

Now it is time to create your "How to" notes. Let's say that your first job function is preparing the daily bank deposit. I have no idea how your dealership prepares a bank deposit – but you do – or the person doing it does. Ask them to explain how they do it in detail and then email it to you. Next, look carefully at the steps. Are they necessary? Do you need to add some step for better Internal Control? Looks at my tips in Tasks for the Daily bank deposit and possible cut and paste that into the document. Once you have it in a Word document, you can cut and paste it into the How To notes. You can also put it all in one Word document. To make it as a training manual, you would first highlight the title **Preparing the Daily Bank Deposit** and from the formatting box (it probably says Normal right now) click and select Heading 2. Continue typing the rest of the function instructions and save your file. Now go to the first page and press Control and Enter to insert a new blank page. From the pull down menu select Insert, Reference, Index and Tables, and click on the tab, Table of Contents. Select a style that you like and click OK. You should have that first topic, "Preparing the Daily Bank Deposit" and the page number. At this point you probably need to add page numbers to your document, so go to the pull down menu again and select Insert, Page Numbers. I realize this seems like a lot to do, but if you start creating the document this way, you'll have a very valuable training manual when you are done. Now you can put this document on a shared folder on your network and create a shortcut to that document with a hyperlink. After you have created the shortcut, right click on it and copy the location. This is the hyperlink you can put in emails to your staff.

Let's say that you now want to start a cross training program. If your Accounts Payable clerk currently does the daily bank deposit, you might want to also have the payroll clerk and new car biller learn it for back up. Before you start sending an email with a link to your document and the chapter you'd like them to study, think of the other items you would like them to learn. For a payroll clerk, you might also want her to learn how to bill out car deals and reconcile account payable statements. Make a full program for them with expected completion dates. If you can combine your functions with a similar training program on your DMS system, that is even better. Create a login for each of them on the training website, then login in under their user ID and add classes to their tasks.

Now that you have created all the job functions in this big document, you can now create some quizzes for your employees. Why? Everyone learns better if they have a measurement of their comprehension. My next suggestion is odd, but I like to have the person doing the live training to give the quiz orally. This is the scenario; you have a payroll clerk that just read the chapter on "Preparing the bank deposit" and now sits one day with the accounts payable clerk and watches her do the bank deposit. When she is done watching, the accounts payable clerk hand the payroll clerk the quiz and they go through it together. There is a small evaluation at the bottom and the trainer either "signs off" the learner on this function or recommends more training. It is important to start documenting your training program. Keep track of what classes have been completed, the quizzes, and amount of training time done by each employee. Obviously you'll determine reviews and advancements on how serious your staff takes this training program that you went to all the trouble to create. How does a training program reduce expenses? Obviously if there are fewer mistakes, then you'll avoid overpaying commissions, technician time, and vendors. You'll also save overtime because instead of a few trained people being able to "cover" for absences, you'll have a wide selection of staff that might need the hours without going into overtime wages. Finally, studies have proven that trained employees have higher job satisfaction, thus you'll retain them longer and have less turnover.

Why Create a Job Description?

One of the first questions I ask when a manager or dealer is disappointed in the performance of one of their employees is, "what does it say in their job description?" I knew a dealer that was furious that the controller had not reconciled the bank statement in over a year – but sure enough that was not in his job description. Many of you might argue that everyone knows that making sure the bank gets reconciled is a basic function of being a controller – but for this poor guy, the bank was reconciled by his CPA firm in his prior job. I've had controllers that didn't know they were supposed to follow up on rebates, reconcile floor plan statements, or even make payroll deposits.

A key to a job description is to make sure these holes don't happen and most importantly, that you get what you want from that person. If you hired me as your controller, your objectives might be for me to optimize your computer system and work with the other managers on their department's profitability. I'd be shocked to find out that you expected me to file, cashier and answer the phone – but I know many controllers that have that as part of their daily functions. Before you hire another manager, grab a copy of their job description and look at what is expected of them. If you don't have any job descriptions yet, then in the next section are some sample job descriptions to get you started. My job description format has four sections; position/department, general description, duties and responsibilities, and skills/physical demands. Let's look at how each of these sections helps you prevent disappointments and make more money.

The "Reports to:" is more important than you think. It clearly defines the department that the person works for and who they report to. I can't tell you how many times in my career that I didn't really know who I worked for. When I was an assistant to the F&I department, I had six or more bosses; the F&I managers, the sales managers, the general manager, the F&I director and of course the controller got stuck with me since I was a female. I know I didn't make effective use of my time since I was often used to fetch a sandwich for the sales managers. I like to think of the general description as a mission statement for the employee and is best when it is three sentences or less. It is the Duties and Responsible section that should make you money with this employee. What are your objectives for this position? For example in my job description for an Internet manager, item six is, "Review billings from web site providers and reconcile any leads or development costs to actual logs." This expense cutting function normally saves the dealership thousands of dollars. Don't expect that your Internet manager is doing this automatically. The last section that defines skills and physical demands is the one that protects you legally. How can you fire someone for a bad driving record if it was never defined that their job required them to take the deposit to the bank or perform dealer trades? If you need this person to drive a company vehicle, then make sure you include that requirement. If they need to be able to lift file boxes, then make sure they can pick up 50 pounds if that is part of their job. You might want to run that section by your legal advisor just to make sure you have the correct requirements or wording for the position. No matter what format you use for a job description make sure you create one for every employee – and make sure the people involved get a copy. If you hire a service advisor – give them a copy and give a copy to the service manager. I'd prefer the service manager edit the description before you even run the advertisement for the position. As you can imagine, I have added to the F&I manager job description the last item; "Submit the contracts to each lender with all the proper documentation to ensure prompt payment and maintain a record of when and how contracts were submitted."

Accounts Payable Clerk

Reports to: Office Manager or Controller

General Description:
The Accounts Payable Clerk is responsible for checking the authorization of payments, creating accounts payable checks, and reconciliation of payment schedules and vendor statements.

Duties and Responsibilities:

- ❒ Compare purchase orders to invoices, checking price extensions, and sales tax calculations.
- ❒ Enter invoices into the computer system from vendors, checking that the payments have been authorized.
- ❒ Maintain the authorized vendor list, periodically checking for the best prices.
- ❒ Reconcile vendor statements to invoices posted and authorized for payment.
- ❒ Investigate any invoices on the vendor statements that are missing.
- ❒ Reconcile any accounts on the accounts payable schedule that cannot be verified by a vendor statement or invoices.
- ❒ Reconcile the sublet schedule and investigate any charges that are not work-in-process or overaged.
- ❒ Provide management with accounts payable estimates for cash flow purposes.
- ❒ Keep a record of forms supply; a copy of the form and who it was last ordered from.

Skills Required and Physical Demands:

1. Excellent math and verbal skills. These skills may be tested and evaluated.
2. At least one year accounting experience.

Human Resource Manager

Reports to: Controller or CFO

General Description:
The Human Resource Manager is responsible for employee relations and conformity with state and federal employment regulations.

Duties and Responsibilities:

- ❐ Recruit new employees by placing advertisement as required by department managers.
- ❐ Interview potential employees, administer screening tests, and provide reference checks.
- ❐ Maintain database of interviews and new hires to meet government requirements.
- ❐ Provide new employee orientation, including completion of all required paperwork and dealership tour.
- ❐ Handle all terminations including exit interview and required documentation.
- ❐ Maintain all employee benefit programs including competitive bids for medical and dental insurance.
- ❐ Update employee hand book as necessary and review all company memos to employees.

Skills Required and Physical Demands:

1. Previous experience as a Human Resource Manager or College Degree/Certification in Human Resources.
2. Ability to communicate with employees and management.

Office Manager

Reports to: Controller, CFO, or Dealer

General Description:
The Office Manager is responsible for the supervision of the office staff, internal controls, and operational reporting.

Duties and Responsibilities:

- ❒ Supervise the daily flow of accounting data from the operational departments into the computer system.
- ❒ Report the status of cash balances and cash requirements to the controller, dealer, or CFO.
- ❒ Schedule the office staff in a manner to provide the best coverage of cashiering and phones for the operational departments.
- ❒ Provide a daily report of operational profit (MIS or DOC) and monthly reports of financial condition (financial statements, GL, schedules).
- ❒ Oversee the computer system and provide ongoing training for the office staff.
- ❒ Report to the controller, dealer, or CFO any variances in gross profit or expenses that are out of guidelines.
- ❒ Reconcile bank accounts and all balance sheets accounts requested by management.
- ❒ Supervise and approve the purchasing of office supplies, forms, and other supplies required by the office and operating departments at the lowest cost.

Skills Required and Physical Demands:

1. Associate degree in accounting and/or equal amount of experience.
2. Ability to communicate with fellow managers and prepare written financial reports.

Accounts Receivable Clerk

Reports to: Office Manager or Controller

General Description:
The Accounts Receivable Clerk is responsible for maintaining the accounts receivable schedule, customer accounts, and collection status.

Duties and Responsibilities:

- ❐ Apply payments for accounts receivable customers, noting any shortages.
- ❐ Send copies of invoices for any invoices not paid or call customers about variances.
- ❐ Notify management of any collection difficulties, making copies of all correspondence and charges for the responsible manager to contact the customer
- ❐ Reconcile any accounts that pay by invoice instead of statement (if on balance forward AR).
- ❐ Provide new account forms to department managers or customers upon request.
- ❐ Verify references on new account forms and forward completed application to management for approval.
- ❐ Send out billing statements monthly and print an aged trial balance.
- ❐ Send out past due statements for any past due accounts.
- ❐ Periodically (weekly) have a meeting with the department managers to review any past due accounts.
- ❐ Provide credit history files for account status inquires per management's guidelines.
- ❐ Collect NSF checks by verifying funds, exchanging for cashiers checks, or making a copy for the departmental manager.
- ❐ Insure that the dealership is in compliance with current collection regulations regarding the hours to call, place of employment calls, and number of contacts allowed.

Skills Required and Physical Demands:

1. Ability to communicate well over the phone and in written correspondence.
2. Excellent computer and math skills. These skills may be tested and evaluated.
3. Previous customer service, collection, or accounting experience.

Vehicle Biller

Reports to: Office Manager or Controller

General Description:
The vehicle biller is responsible for the flow of car deals, vehicle receivables and contracts, sales commissions, and titling through the accounting office.

Duties and Responsibilities:

- ❒ Provide the controller with a daily listing of contracts and vehicle receivables due and proceeds.
- ❒ Finalize the vehicle deals and reconcile the office gross profit to the sales department gross profit.
- ❒ Run daily reports from the F&I systems of total deals, reconciling it to the office completed deals for a "deals missing" listing.
- ❒ Review any titling problems with the responsible managers and provide a monthly titling "heat sheet" for missing documentation.
- ❒ Approve the final sales payroll commission summaries verifying held deals and voucher amounts.
- ❒ Provide the controller with final sales count figures for the financial statement memo and inventory listings of over aged vehicles.
- ❒ Maintain the computer interface between sales, F&I, and accounting for error free transfer of information between departments.
- ❒ Supervise the collection of incentives and rebates, maintaining a master file of program guidelines.
- ❒ Create a daily heat sheet of all monies and titling issues

Skills Required and Physical Demands:

1. Previous experience as a biller and title clerk.
2. Excellent math and communication skills.

Chief Financial Officer

Reports to: Dealer or CEO

General Description:
The Chief Financial Officer is responsible for the financial management of the dealership group with a focus on profitability, internal controls, and asset management.

Duties and Responsibilities:

- ❐ Establish accounting guidelines and policies for uniformity of financial reporting between dealerships.
- ❐ Prepare consolidated financial analysis reports comparing the various dealerships performance with each other.
- ❐ Establish metrics and performance goals for each dealership and department.
- ❐ Present regular financial and policy meetings for the General Managers and Controllers to discuss common problems and goals.
- ❐ Create an internal control policy and if possible an internal control staff to monitor the systems and processes at each dealership and compliance with government regulations.
- ❐ Supervise the relationships between each dealership and their manufacturer, lenders, and computer vendors, creating volume discount agreements whenever possible.
- ❐ Supervise the selection and training of controllers/office managers.
- ❐ Arrange for insurance coverage, employee benefits, legal and audit services, and any other services that protect the company's interest and provide volume discounts.
- ❐ Prepare an annual business plan that forecasts cash requirements, profitability, major asset purchases, and personnel changes.
- ❐ Determine each dealership's computer utilization level and set computer security policies.

Skills Required and Physical Demands:

1. Degree in accounting or equal amount of experience.
2. Former Controller or Office Manager in automotive or related industry.
3. Ability to communicate with key managers and prepare financial reports.

System Administrator

Reports to: Controller or GM

General Description:
The System Administrator provides maintenance and support of various DMS, CRM, and other software programs, the dealership network, and user PCs.

Duties and Responsibilities:

- ❒ Daily backup of data and maintenance of disk space with performance information.
- ❒ Installation of software updates and maintaining guidelines for databases and field entries.
- ❒ Changing form design as necessary.
- ❒ Replacing defective equipment with spare equipment and coordination of bids for new purchases.
- ❒ Calling vendor support when necessary.
- ❒ Providing reports to managers and reformatting of forms and letters.
- ❒ Training of new personnel on software and attendance at update classes.
- ❒ Creation of a technology budget and long-term planning for technology
- ❒ Protection of the system computers from viruses and attacks
- ❒ Maintenance of email system
- ❒ Update of software programs to current versions
- ❒ Assist management in the selection of technology
- ❒ Restriction of Internet access as requested by management
- ❒ Security control of passwords and network access
- ❒ Safeguard of the customer database and other company data

Skills Required and Physical Demands:

1. Former automotive system administrator or vendor experience (1 year minimum) or technical degree in computer science.
2. Ability to lift and carry 50 pounds.

Payroll Administrator

Reports to: Office Manager or Controller

General Description:
The Payroll Administrator is responsible for processing the payroll cycles, maintaining the employee payroll files, and submitting tax payments and reports.

Duties and Responsibilities:

- ❒ Check timecards, payroll summaries, and bonus calculations for accuracy.
- ❒ Enter payroll data into the in-house computer system or transmit to outside payroll processor.
- ❒ Enter any deductions for garnishments, insurance, uniforms, employee receivables, 401K, or other requested deductions.
- ❒ Process the payroll cycles and print the employee paychecks.
- ❒ Maintain both the computer employee record file and the physical employee document file.
- ❒ Make any required tax deposits and other deduction payments to third parties and taxes, complete the necessary monthly, quarterly, and annual tax and summary reports for payroll taxes, workers compensation, insurance, 401K plans, and garnishments.
- ❒ Order payroll checks, W-2 forms, employee applications, and other payroll forms as necessary.
- ❒ If there is no human resource manager, perform those job duties
- ❒ Determine proper accounting of salary and tax expenses and accruals and maintain the accounting interface or post the payroll summaries.
- ❒ Provide management with payroll estimates for cash flow purposes.

Skills Required and Physical Demands:

1. Excellent math and verbal skills. These skills may be tested and evaluated.
2. At least one year accounting experience.

Collection Manager

Reports to: Controller

General Description:
The Collection Manager is responsible for daily collection of all past due trade accounts, NSF checks, and if necessary buy/here/pay/here payments.

Duties and Responsibilities:
- ❒ Provide controller and general manager with weekly status reports of all accounts and collection status.
- ❒ Make daily collection calls and contacts with past due accounts.
- ❒ Maintain policies for customer contacts within federal and state regulations for the hours to call, place of employment calls, and number of contacts allowed.
- ❒ Hold regular collection meetings with operational managers to discuss credit policies and past due accounts.
- ❒ Recommend new accounts for credit limits based on investigation of credit applications.
- ❒ Collect NSF checks by verifying funds, exchanging for cashier checks, or collecting cash from the customers.
- ❒ Update computer records for past due accounts restricting future credit purchases.

Skills Required and Physical Demands:
1. Excellent communication skills.
2. Ability to sit for long periods of time for phone contacts.
3. Previous collection or customer service experience.

Purchasing Agent or Buyer

Reports to: Controller or GM

General Description:
The Purchasing Agent or Buyer is responsible for saving the dealership money by negotiating the best prices for office and shop supplies, outside services, and sublet repairs. Authorizes all purchases and checks the purchase orders against billing invoices for accuracy.

Duties and Responsibilities:

- ❒ Keep a sample of all forms currently used by the dealership. Track the usage of forms, current inventory, and ordering history.
- ❒ Negotiate the purchase of printed forms, avoiding rush and excess shipping charges.
- ❒ Maintain the inventory of office supplies and the vendors to order from for the best price.
- ❒ Keep a listing of all sublet repair vendors, the services that they perform, and the agreed upon price. Actively search for new vendors at better prices.
- ❒ Issue purchase orders, noting quantity, description, agreed upon price, accounting information, and if necessary authorizing signature.
- ❒ Match purchase orders to invoices, checking for billing errors and discounts. Verify correct sales tax rates and freight charges. Challenge any excessive shipping or handling charges.
- ❒ Track advertising costs. Daily notify managers if they have exceeded or will exceed their monthly budget. Compare media billings to contract rates and tear sheets for accuracy.
- ❒ Keep a file for all outside services including existing contract, expiration date, and agreed upon rate. Negotiate all bids for the best prices for cleaning services, uniform services, trash, and other outside services. Check all monthly billings against contracts.
- ❒ Provide suggestions on internal control of inventory and usage of shop supplies.
- ❒ Suggest recycling and alternate products.
- ❒ Maintain the vendor listing for suppliers and best prices.

Skills Required and Physical Demands:
1. Previous experience as a purchasing agent, parts inventory clerk, or accounts payable clerk.
2. Ability to lift 50 pounds.

Title Clerk

Reports to: Office Manager or Controller

General Description:
The Title Clerk is responsible for processing vehicle title applications, maintaining the title for in-stock vehicles, and purifying the title fee schedule.

Duties and Responsibilities:

- ❒ Verify that funds have been collected and the correct lien holder paid off before processing title applications.
- ❒ Check for accuracy in the application and complete any missing items.
- ❒ Keep management and the sales department informed of all title regulations.
- ❒ Maintain either a documentation manual or computer inquiry system to verify out-of-state titles.
- ❒ Perform a monthly "title to vehicle" inventory check and report to management any missing titles and their status.
- ❒ Prepare a monthly report to management of any funds not collected from wholesalers due to missing or incomplete title work.
- ❒ Purify the title fees schedule and provide management with a listing of any uncollected fee.
- ❒ Refund any excess fees according to management's guidelines.
- ❒ Provide periodic training sessions to F&I managers and sales personnel about title regulations and procedures. Issue written memo whenever regulations change.
- ❒ Attend seminars held by local licensing bureaus and any available training meetings on title regulations.

Skills Required and Physical Demands:

1. Excellent math and verbal skills. These skills may be tested and evaluated.
2. At least one year accounting or licensing agency experience.

Super Controller

Reports to: Dealer, CFO or GM

General Description:
The Controller is responsible for the financial management of the dealership with a focus on profitability, internal controls, and asset management.

Duties and Responsibilities:

- ❒ Preparation of a monthly financial statement within 3-5 working days according to the company's accounting guidelines and each manufacturer's format.
- ❒ Analysis of cash flow, including forecasting of cash needs and transfer of excess cash into an interest bearing account (or paying off floor plan).
- ❒ Supervision of compliance with government regulations including forms 8300 and 1099, payroll taxes and deposits, income taxes, 401K plans, sales tax, hazardous materials, and employee safety.
- ❒ Oversee all collection efforts and report any past due accounts to the dealer and department managers.
- ❒ Supervise the relationship with the manufacturer including warranty claims, rebates, interest protection, and coop advertising. Verify that all deliveries are reported accurately.
- ❒ Provide department managers with a detail of expenses, gross profit analyses, employee productivity reports, and any other information they need to manage their departments.
- ❒ Oversee the computer utilization and training for each department and password security.
- ❒ Review the detail of all general ledger and scheduled accounts and analyze the prompt conversion of key asset accounts into cash.
- ❒ Prepare monthly forecasts and budgets with department managers and follow up with forecast to actual reports.
- ❒ Review the reconciliation of general ledger accounts with outside bank sources; bank reconciliation, reserve accounts, factory payable, and floor plan payable.
- ❒ Supervise the preparation of all information for the outside CPA firm with an objective to minimize the audit or review costs.

Skills Required and Physical Demands:

1. Degree in accounting or equal amount of experience.
2. Ability to communicate with fellow managers and prepare written financial reports.

Customer Contact Manager

Reports to: General Manager

General Description:
The Customer Contact Manager is responsible for all contacts with customer from the initial "up" tracking through sales and service follow-up, and eventually with CRM and service reminders.

Duties and Responsibilities:

- ❒ Maintain the sales department's lead or CRM software, phones, computers, and supplies.
- ❒ Assist salespeople with all customer contacts including Internet, phone, and walk-in leads.
- ❒ Maintain the website including the purification and uploading of the vehicle inventory, creation of specials, contact information.
- ❒ Supervise the entering of call notes by the sales department and flag customers for follow-up
- ❒ Determine with management the next data mining campaign.
- ❒ Create and print reports for management – closing ratios, salesperson analysis, CSI scores by salesperson/advisor, campaign profitability analysis.
- ❒ Prevent duplicate customers and periodically clean the databases of duplicates.
- ❒ Enter change of addresses or use www.cleanlist.com before each mailer.
- ❒ Create other custom reports and labels as necessary for customer retention or contact.
- ❒ Calculate monthly the dealership's customer retention rate for sales & service
- ❒ Provide **Daily Work Plans** to salespeople.
- ❒ Create and supervise a service reminder system.
- ❒ Account for all marketing and CRM costs including the purification of lead generator billings, CRM software, direct mailings, etc.

Skills Required and Physical Demands:
1. Excellent math, technology and verbal skills. These skills may be tested and evaluated.
2. Previous salesperson, sales manager, Internet manager, CSI staff, F&I manager, system administrator or accounting clerk experience.

Assistant Manager/Closer

Reports to: Sales Manager

General Description:
The Assistant Sales Managers or "closers" are responsible for monitoring and completing the sales of their team members.

Duties and Responsibilities:

- ❒ Supervise team members paying special attention to the flow of customers from "ups" to write-up.
- ❒ Provide or arrange any training necessary on paperwork requirements and product knowledge.
- ❒ Help team members "close" deals and work as a liaison between team members and the sales and F&I managers.
- ❒ Monitor demonstration rides, insuring the safety of team members.
- ❒ Control keys and dealer plates and assist team members in completing any borrowed car agreements.
- ❒ Help team members with their customer retention program and provide a daily status report to the sales managers.
- ❒ Work closely with the F&I office to collect any missing paperwork or past due funds.
- ❒ Make inventory purchase suggestions to the sales managers based on customer and team members' requests.
- ❒ Help sales managers control gas expense and lot damage.
- ❒ Maintain a record of each team member's attendance notifying management when absences have become excessive.

Skills Required and Physical Demands:

1. Previous salesperson or F&I manager experience.
2. Ability to communicate with customers, other employees, and managers with verbal and written reports.
3. Ability to be insured to drive a company owned vehicle.

BDC/CRM Manager

Department: Sales

Position reports to: General Manager or General Sales Manager

General Description:
The BDC/CRM Manager's goal is to promote long term customer retention by either assisting the sales force or customer retention staff in a continuous contact strategy.

Duties and Responsibilities:

- ❒ Maintain the BDC/CRM or BVD office equipment; phones, computers, and supplies.
- ❒ Assist salespeople or BDC/CRM staff with customer contacts through training and data management.
- ❒ Provide a report to the salespeople or BDC/CRM staff of all customers who have not been in for service for the past 4 months.
- ❒ Provide a report of customers who are coming in for service that did not buy a vehicle at the dealership.
- ❒ Provide a report of closed repair orders each month or day by salesperson.
- ❒ Keep track of the first service appointment for customers and provide reminders.
- ❒ Provide a report of total customer pay and customer parts sales sorted by salespersons' number.
- ❒ Track the results either by salespersons' or BDC/CRM staffs' results of their customer contacts and update the customer database.
- ❒ Create other custom reports and labels as necessary for customer retention or contact.
- ❒ Calculate monthly the dealership's customer retention rate for sales & service departments.

Skills Required and Physical Demands:
1. Excellent math and verbal skills. These skills may be tested and evaluated.
2. Previous salesperson, sales manager, F&I manager, system administrator or accounting clerk experience.

Customer Service Manager

General Manager or Service Manager

General Description:

The Customer Service Manager is responsible for maintaining high customer retention, customer satisfaction, and monitoring customer complaints.

Duties and Responsibilities:

- ❒ Oversee the new vehicle delivery process, watching for quality control in the vehicle condition and sales person's knowledge.
- ❒ Monitor that sales customers are coming in for service after delivery by having the first appointment made by the sales person and provide listings of any customers who have not come in for service.
- ❒ Contact service customers after their visit to monitor quality control and satisfaction.
- ❒ Keep records of all factory customer satisfaction indexes and reports.
- ❒ Provide listings of all customers who are not coming back for service.
- ❒ Monitor comeback reports of repeat visits and investigate problem areas.
- ❒ Keep records of all customer complaints and solutions.
- ❒ Provide training to employees on better customer service.
- ❒ Contact sales customers after delivery to monitor customer satisfaction.
- ❒ Monitor continuous improvement efforts that will increase customer satisfaction and customer loyalty.

Skills Required and Physical Demands:

1. Ability to communicate well over the phone and in written correspondence.
2. Previous sales or customer service experience.

F&I Assistant

Reports to: Controller, F&I Manager, GM or GSM

General Description:
The F&I Assistant helps the F&I Managers (Business Managers) complete the deal paperwork, maintain logs, and updates the computer system.

Duties and Responsibilities:

- ❒ Control the status of deals; pending, closed, and sent to accounting.
- ❒ Update the computer system as each status changes.
- ❒ Assist the managers in completing the deal paperwork, including preparing the contracts to send to the bank or captive finance company.
- ❒ Maintain an F&I income log (hopefully on the computer system) and reconcile the F&I income to the accounting office balances.
- ❒ Verify payoffs, insurance coverage, and delivery status of the deals.
- ❒ Prepare cancellation forms for extended warranty and insurance coverage and respond to customer's inquiries regarding refunds.
- ❒ If necessary, enter fleet and wholesale deals into the computer system to update inventory records and service history.
- ❒ Maintain a rebate manual that outlines all current rebate and incentive programs for each manufacturer.
- ❒ Assist the managers in the collection of the "heat sheet" items, such as difference in payoffs, NSF checks, past due payments, and trade-in titles.
- ❒ Run a daily report of "total deals" and reconcile it to the amounts of deals shown by the sales office and accounting office.

Skills Required and Physical Demands:

1. Previous experience as an automotive accounting or title clerk.
2. Excellent math and communication skills.

F&I Manager

Reports to: General Sales Manager or General Manager

General Description:
The Finance and Insurance Manager (or Business Manager) has the overall responsibility to finalize the financial and insurance portion of the vehicle sales transaction.

Duties and Responsibilities:

- ❒ Submit credit applications for pending deals to financial institutions for approval according to management's guidelines.
- ❒ Strive to obtain the lowest buy-rate (cost) for financing, paying careful attention to penetration quotas necessary for discounts on floor plan rate.
- ❒ Maintain a current folder or binder containing all extended warranty policy costs and guidelines, insurance rates, and financing buy-rates.
- ❒ Greet sold customers and invite them into the F&I office to finalize their paperwork .
- ❒ Prepare all necessary paperwork paying careful attention to completeness and conformity with local, state, and federal regulations.
- ❒ Collect any monies due, or verify funds with the accounting office.
- ❒ Maintain all state licenses required to sell insurance and extended warranty policies.
- ❒ Train the sales force on how to prepare customers for F&I.
- ❒ Keep a log (preferably on the computer system) of all deals completed in finance, deals in accounting, and pending deals with their status.
- ❒ Prepare a daily report (preferably on the computer system) of all unit totals, gross and F&I profit totals.
- ❒ Maintain the F&I forms library, making sure that the most current and accurate forms are being used.
- ❒ Submit the contracts to each lender with all the proper documentation to ensure prompt payment and maintain a record of when and how contracts were submitted.

Skills Required and Physical Demands:

1. Excellent verbal, math, and leadership skills. These skills may be tested and evaluated.
2. Prior sales experience, accounting office contract experience, F&I assistant, or loan officer.
3. Ability to be insured to operate a motor vehicle.

Fleet Manager

Reports to: General Sales Manager

General Description:
The Fleet Manager is a specialized sales person who sells to business clients, large major accounts with more than 2 purchases per year, dealership employees, and factory personnel.

Duties and Responsibilities:

- ❒ Contact business clients from a qualified prospect list through phone calls, client's visits, or personal letter.
- ❒ Prepare quotes and bids for commercial clients according to management's pricing guidelines.
- ❒ Maintain the client database according to management's guidelines, either on the existing in-house computer or specialized sales prospecting software. Process regular updates from database provider.
- ❒ Negotiate final agreed upon price with client and management.
- ❒ Prepare and process required paperwork according to management's policies.
- ❒ Deliver vehicles to clients and collect funds.
- ❒ Establish service department accounts and arrange first appointment for clients.

Skills Required and Physical Demands:

1. Excellent math and verbal skills. These skills may be tested and evaluated.
2. Previous sales experience with full knowledge of product line, or ability to attend training.
3. Ability to be insured to operate a motor vehicle.

General Manager

Reports to: Dealer or President

General Description:
The General Manager has the primary responsibility to maximize the net profit off all the departments while maintaining the highest level of customer satisfaction and retention.

Duties and Responsibilities:

- ❒ Hire and supervise departmental managers.
- ❒ Prepare a monthly forecast and budget based on input from the department managers.
- ❒ Periodically review the pay plans of key personnel in all departments to ensure that the dealership's goals are being met.
- ❒ Analyze the detail of expenses each month and work with managers to cut costs.
- ❒ Review the employee safety program, environment regulations, legal issues and insurance guidelines for the departments.
- ❒ Maintain good relations with factory personnel paying special attention to vehicle ordering, advertising programs, and customer satisfaction scores.
- ❒ Open the mail daily and review any questionable items with managers. Perform a monthly "checks in the mail" audit.

Skills Required and Physical Demands:

1. Previous GSM, Controller, CFO or related management experience.
2. Ability to communicate with customers, departmental employees and other managers with verbal meetings and written reports.

General Sales Manager

Reports to: Dealer or General Manager

General Description:
The General Sales Manager has the overall responsibility of both new and used car departments, their respective managers, their functions, sometimes the used car buyers, and acts as liaison to the factory representatives.

Duties and Responsibilities:

- ❒ Develop, delegate implementation of, and maintain dealership sales system, consistent with philosophy of dealer or General Manager.
- ❒ Communicate compensation plans to sales managers, and help them maintain their focus' on variable gross, and concurrently variable expenses.
- ❒ Maintain clear communication between sales department managers.
- ❒ Establish, maintain, and merchandise proper mix of new cars and trucks, and rapid rate turning used car inventories.
- ❒ Develop focused sales promotions for dealership based on factory incentives, special auction purchases, or seasonal changes in the marketplace.
- ❒ Act as sounding board for concerns sales managers may have that are not resolvable within their limits of power.
- ❒ Monitor, maintain, and exceed manufacturers' expectations of market penetration.
- ❒ Assist sales managers and sales staff in maintaining a continuous customer contact strategy.
- ❒ Monitor prospect/customer impressions of dealership through either surveys, or mystery shoppers, frequently.
- ❒ Recognize, and resolve all sales problems with the dealer's philosophy in mind.
- ❒ Report to the General Manager any problems, irregularities, or suspicions discovered through day-to-day observation.

Skills Required and Physical Demands:

1. Excellent communication and presentation skills, both written and verbal.
2. Good reasoning skills that mirror the philosophy of the dealer.
3. Successful managing of sales persons, in an advisory capacity.
4. Complete understanding of the various departments, and their functions in a dealership.

Get Ready Manager

Reports to: Sales or Service Manager

General Description:
The Get Ready Manager has the responsibility to monitor the flow of vehicles from the factory to the sales lot and from the lot to customer delivery.

Duties and Responsibilities:

- ❒ Supervise any get ready technicians to maximize productivity and reduce department expense.
- ❒ Control the get ready supplies by taking a monthly inventory and working with vendors for the lowest prices.
- ❒ Provide the accounting office with the proper documentation when receiving new or auction vehicles.
- ❒ Maintain the vehicle database on the web site to insure accuracy and up-to-date product.
- ❒ Make sure the warranty administrator files claims for any transportation damage or missing equipment.
- ❒ Obtain daily lists of vehicles being delivered to ensure that customer's vehicles are ready for pickup and in excellent condition.
- ❒ Verify that each new vehicle has a full tank of gas (if it is a dealership or factory policy) and provide a copy of the gas invoice to the accounting department.
- ❒ Promptly close all internal get ready repair orders and run a daily listing of any open repair orders for management.
- ❒ Control the get ready shop tool and supply inventory to prevent theft and loss.
- ❒ Understand all factory and dealership policies regarding PDI requirements and dealer trades.
- ❒ Monitor any ordered parts for new vehicles in stock.
- ❒ Inform the sales department of any missing factory or dealer stickers.

Skills Required and Physical Demands:

1. Previous technician, service advisor, or related management experience.
2. Ability to communicate with technicians, managers, and customers both verbally and with written repair orders.
3. Ability to be insured to operate a motor vehicle.

Internet Manager

Reports to: Sales Manager or GM

General Description:
The Internet Manager is responsible for all communication between Internet customers and other dealership personnel, as well as maintenance of the Internet site.

Duties and Responsibilities:

- ❒ Every 15 minutes retrieve all email for the dealership site and route to appropriate personnel keeping a record of all contacts.
- ❒ Respond to sales inquires according to management's guidelines providing pricing and specification information.
- ❒ Maintain the vehicle database on the web site to insure accuracy and up-to-date product information.
- ❒ Regularly review dealership information on the web site and provide the webmaster with any corrections to personnel, hours, address, etc.
- ❒ Provide a monthly report to management that contains web contacts, outcome of the contact, and other requested information.
- ❒ Review billing from web site provider and reconcile any leads or development costs to actual logs.
- ❒ If authorized by management, create a service appointment for any Internet customers that request it and provide an email reminder.
- ❒ Train salespeople on using the Internet for vehicle inventory searches and comparisons.

Skills Required and Physical Demands:

1. Excellent computer skills, including experience with in-house system, email, and Internet products.
2. Ability to communicate verbally and especially with written communications.
3. Previous sales or customer service experience.

Inventory Administrator

Reports to: Sales Manager

General Description:
The Inventory Administrator provides management with information on the current inventory status, on order vehicles, and over-age inventory.

Duties and Responsibilities:
- ❒ Provide a weekly inventory report to sales managers, controller, general manager, and dealer. The report should include an aging summary, days supply, on-order vehicles, and sales history.
- ❒ Create suggested orders according to manufacturers ordering schedules to review with management and obtain approval.
- ❒ Process orders to manufacturers and monitor built statuses.
- ❒ Monitor the receipt of vehicles and be responsible for minimizing the time from receipt to available for sell. Open internals for safety checks and PDIs and coordinate with the service department for the completion of these items.
- ❒ Maintain the computerized database of the inventory including the additions of equipment, accuracy of statuses, and sales reporting. Provide a monthly computer to physical inventory report and computer to manufacturer purification.
- ❒ Keep a control over keys, owners manuals, and the condition of the vehicles.
- ❒ Coordinate the repair of transportation damage, lot damage, and replacement of manuals and keys.
- ❒ Maintain a record of customer orders and status.

Skills Required and Physical Demands:
1. Former salesperson, sales manager, accounting staff member, or other related automotive experience.
2. Computer experience or technical school computer education.
3. Ability to perform a physical inventory of vehicles.
4. Insurable driving record.

Sales Manager

Reports to: Dealer, General Manager, or GSM

General Description:
The Sales Manager has overall control over the sales department activities, including supervision of the sales force and management of the inventory.

Duties and Responsibilities:

- ❐ Supervise the sales department staff paying special attention to the flow of customers to the salespeople including "ups", appointments, phone inquires, and general showroom control.
- ❐ Maintain internal control over the new and used vehicle inventory to prevent theft, vandalism, and lot damage.
- ❐ Develop marketing strategies to increase sales and gross profit using advertising, promotions, and sales incentives.
- ❐ Approve new vehicle orders and auctions purchases within management's guidelines.
- ❐ Develop a customer retention program that includes a follow-up of prospects and a flow of sold customers to the service department.
- ❐ Control purchase orders for sales department expenses paying special attention to gas expense, reconditioning, and advertising.
- ❐ Assist accounting office with the collection of contracts in transit, vehicle receivable and NSF check accounts.
- ❐ Arrange for insurance coverage, employee benefits, legal and audit services, and any other services that protect the company's interest and provide volume discounts.
- ❐ Assist with the management of the finance and insurance office especially with the relationship with lenders.
- ❐ Recruit, hire, and train the sales force and sales support staff.

Skills Required and Physical Demands:

1. Prior experience as a sales person, F&I manager, or related industry manager.
2. Excellent math, verbal, and leadership skills. These skills may be tested and evaluated.
3. Ability to be insured to operate a motor vehicle.

Salesperson

Reports to: Sales Manager

General Description:
The Salesperson has the primary responsibility to follow the steps that management has created to result in a successful vehicle sale and long term customer retention.

Duties and Responsibilities:

- ❒ Make regular customer contacts by the means outlined by management such as floor time, phone "ups", mail, and personal visits.
- ❒ Maintain a record of all customer contacts and regularly report the progress to the sales manager.
- ❒ Obtain the training necessary for factory qualifications and the product knowledge for each model.
- ❒ Perform regular and scheduled contacts with previous customers and report the performance to the sales manager.
- ❒ Take test-drives with customers according to management's guidelines for personal safety.
- ❒ Present to management all offers made by customers to purchase vehicles.
- ❒ Prepare and process the required paperwork to finalize the vehicle sale.
- ❒ Deliver vehicles to customers, collect any monies due, and process the trade-in as requested by management.
- ❒ Arrange the first service appointment for customers and contact the customer before and after the first appointment.

Skills Required and Physical Demands:

1. Excellent verbal and math skills. These skills may be tested and evaluated.
2. Previous sales experience with either full product knowledge or the ability to attend training.
3. Ability to be insured to operate a motor vehicle.

Used Car Buyer

Reports to: Sales Manager

General Description:
The Used Car Buyer provides the used vehicle sales department with a supply of Marketable used vehicles within management's financial guidelines and purchasing policies.

Duties and Responsibilities:

- ❒ Review all auction notices and discuss proposed travel expenses and attendance with management.
- ❒ Discuss purchasing requirements and forecast sales for future months with used vehicle managers prior to auction attendance.
- ❒ Obtain an authorization from management for total amount of purchases prior to auction attendance.
- ❒ Attend auctions approved by management and purchase vehicles.
- ❒ Approve all paperwork from auctions and provide proper documentation to the office for purchases.
- ❒ Notify the used vehicle department and reconditioning manager of anticipated delivery of purchases. Arrange service appointments if necessary.
- ❒ Keep track of results from auction purchases including comparisons of gross profits from auction vehicles vs. trade-ins.
- ❒ Develop a method to determine the best purchasing mix based on sales history and local area trends.
- ❒ Keep track of customer interest on vehicles not in stock (lost sales).

Skills Required and Physical Demands:

1. Former used vehicle manager, reconditioning manager, or salesperson with used vehicle knowledge.
2. Ability to travel extensively and inspect used vehicles.
3. Insurable driving record.

Parts, Service and Body Shop – Job Descriptions

Service Advisor

Reports to: Service Manager

General Description:
The Service Advisor has the primary responsibility to sell service labor and parts to customers and promote long term customer retention.

Duties and Responsibilities:

- ❒ Greet customers and carefully record the customer's concerns to generate an accurate work order.
- ❒ Track the status of open repair orders during the day keeping both the customers and dispatcher updated with approvals and delays.
- ❒ Obtain the training necessary to know what is covered under warranty for each franchise.
- ❒ Perform regular and scheduled contacts with previous customers and make appointments for maintenance items.
- ❒ Make sure all customers have been notified when ordered parts arrive and schedule an Appointment to have the work done.
- ❒ Help the cashier close tickets and notify customers that their vehicles are ready for pickup.
- ❒ Explain repair order charges to customers and promptly notify the service manager of any customer service problems.
- ❒ Strive to meet management's metrics for sales per repair order, total gross profit, repair orders per day, and customer satisfaction scores.

Skills Required and Physical Demands:

1. Excellent verbal and math skills. These skills may be tested and evaluated.
2. Previous technician, cashier, warranty administrator, or other advisor experience.
3. Ability to be insured to operate a motor vehicle.

Service Cashier

Reports to: Service Manager or Controller

General Description:
The Service Cashier has the primary responsibility for closing service repair orders, collecting payment, and ensuring customer satisfaction.

Duties and Responsibilities:

- ❒ Close all completed repair orders and notify the service advisor or customer of the total charges.
- ❒ Check the accuracy of the repair orders making sure that the correct amount of labor has been charged and corresponding parts have been entered on the repair order.
- ❒ Maintain the proper procedures for the collection of shop supplies, gas, oil and grease, and sales taxes.
- ❒ Prepare the proper billing information for charge and extended warranty repair orders including necessary copies and purchase orders.
- ❒ Maintain the filing system of repair order copies for quick retrieval of information.
- ❒ Reconcile daily the total of funds collected to the balance in the cash drawer, documenting any discrepancies.
- ❒ Ensure that each customer has been promptly assisted and notify management of any customers that might be dissatisfied with their bill or service work performed.
- ❒ If necessary, assist service advisors, the service manager, and receptionist in taking customer appointments or messages.

Skills Required and Physical Demands:

1. Excellent math and clerical skills.
2. Experience as a cashier, customer service representative, or accounting clerk.
3. Ability to communicate verbally with customers and other employees.

Technician

Reports to: Service Manager

General Description:
The Technician is responsible for performing repair and service on customer and dealership vehicles within management's guidelines for productivity, safety, and efficiency.

Duties and Responsibilities:

- ❒ Receive next repair or service job from dispatcher, advisor, team leader or electronic technician terminal.
- ❒ Perform a diagnosis and report back to management with work that is required and await authorization.
- ❒ Review vehicle history file and any factory bulletins that apply to vehicle make and model, paying special attention to outstanding campaigns.
- ❒ Make notes regarding proposed and additional work on hard copy or technician terminal.
- ❒ After authorization, request any parts required from the parts department and return any parts that have been requested to be saved by the customer or factory.
- ❒ Help team members with their customer retention program and provide a daily status report.
- ❒ After parts are received, complete required authorized repairs and create notes on the hard copy or technician terminal.
- ❒ If required, perform test drive and restore vehicle to original condition making sure that it is clean.
- ❒ Inform dispatcher, advisor, team leader or electronic technician terminal of the job completion and make any final notes regarding suggested additional work.
- ❒ Keep informed of all safety guidelines of management and other agencies, especially the disposal of hazardous waste.
- ❒ Take all required and suggested training, keeping a record of classes and certification completed.
- ❒ Keep the work area clean and maintain the condition of both personal tools and shop tools.

Skills Required and Physical Demands:

1. Ability to lift 50 pounds.
2. Ability to communicate both verbally and with written documentation.
3. Completion of technical training school or factory apprentice program.

Parts Driver

Reports to: Parts Manager

General Description:
The Parts Driver is responsible for the daily pickup and delivery of parts and completion of related paperwork.

Duties and Responsibilities:

- ❒ Load all parts in parts truck for required deliveries.
- ❒ Match part quantities in the truck with parts tickets and provide the manager with a list of deliveries being made and a copy of the tickets being delivered.
- ❒ Create a plan for deliveries using the most effective routes, taking into consideration traffic patterns, customer requirements, and the security of the parts in the truck.
- ❒ Obtain the customer's signature on both the office and customer copy of the parts ticket.
- ❒ Collect funds for any C.O.D. parts – collecting cash instead of checks where indicated.
- ❒ Make frequent calls back to the dealership to check if any parts need to be picked up.
- ❒ After returning to the dealership, reconcile collected funds and signed copies of tickets to delivery records left with the parts manager.
- ❒ Purchase gas for the parts truck as necessary according to management's guidelines.
- ❒ Check oil each time. Advise service manager of any warning lights.
- ❒ Adhere to all safety requirements regarding the use of lift belts and loading aids.

Skills Required and Physical Demands:

1. Ability to lift 70 pounds.
2. Clean driving record and ability to be insured to operate company vehicles.
3. Ability to communicate with customers and fellow employees.

Parts Manager

Reports to: Dealer or GM

General Description:
The Parts Manager has overall control over the physical parts inventory and profitability of the parts department.

Duties and Responsibilities:

- ❐ Supervise the parts department staff to provide the highest level of service to counter, wholesale, and service department customers.
- ❐ Maintain internal controls to prevent shortages and obsolescence. Maintain perpetual inventory checks and annual counts per management guidelines.
- ❐ Develop marketing strategies to increase sales and gross profit using coop advertising funds whenever possible.
- ❐ Approve stock orders and general purchases within management guidelines.
- ❐ Develop goals with the service manager for level of service (over 90%) and stock order performance.
- ❐ Create an approved vendor list optimizing the best purchase discounts and return programs.
- ❐ Assist the accounting office with the collection of past due, COD, and NSF check accounts.
- ❐ Develop pricing matrixes and economic ordering criteria to maximize gross profit without sacrificing sales.

Skills Required and Physical Demands:

1. Prior experience as a parts manager or lead counter person.
2. Excellent math and verbal skills. These skills may be tested and evaluated.
3. Ability to be insured to operate a motor vehicle.
4. Ability to lift 50 pounds.

Fixed Operations Manager

Reports to: Dealer or General Manager

General Description:
The Fixed Operations Manager has the primary responsibility of maximizing the net profit of the service, body, and parts departments while maintaining the highest level of customer satisfaction and retention.

Duties and Responsibilities:

- ❒ Hire, supervise, and train the service, body, and parts department managers.
- ❒ Prepare a monthly forecast and budget for the department managers.
- ❒ Compare actual results to the forecast and budget paying special attention to gross profit, expense, and asset management metrics.
- ❒ Periodically review the pay plans of key personnel in the departments to ensure departmental goals are being met.
- ❒ Eliminate "vertical" departmental thinking and create inter-departmental cooperation through pay plan linkage.
- ❒ Analyze the detail of departmental expenses each month and work with vendors to cut costs.
- ❒ Send out competitive bid invitations for most major expenses and assist the office with creating an approved vendor list.
- ❒ Supervise the employee safety program, environment regulations, and insurance guidelines for the departments.
- ❒ Maintain good relations with factory personnel paying special attention to co-op advertising programs, parts return or ordering programs, warranty reimbursements, and customer satisfaction scores.
- ❒ Review the parts management report with the parts manager each month, especially obsolete parts, parts over 12 months old, and parts with no sales.

Skills Required and Physical Demands:

1. Prior experience as a parts or service manager.
2. Excellent verbal and math skills. These skills may be tested and evaluated.
3. Ability to be insured to operate a motor vehicle.

Parts Counterperson

Reports to: Parts Manager

General Description:
The Parts Counterperson is responsible for providing assistance to customers that want to buy parts and processing the final sales transaction.

Duties and Responsibilities:

- ❒ Find the part number, if necessary, for a parts request using electronic catalogs, books, or microfiche.
- ❒ Make a part number inquiry into the parts inventory for the requested part.
- ❒ Invoice the customer for the part charging the correct price, tax, and miscellaneous charges.
- ❒ Collect the funds for the invoice or make sure the cashier receives payment prior to issuing the part.
- ❒ If the part is not in inventory, locate the part from an approved outside source or give the request to the parts manager to locate.
- ❒ Conscientiously follow the parts department procedures for tracking lost sales.
- ❒ Complete a special order part request for any parts that need to be ordered.
- ❒ Assist the back counterperson or stocking clerk as necessary.
- ❒ Learn any new computer updates that affect the parts invoicing/inventory system.
- ❒ Notify customers when special order parts arrive.
- ❒ Adhere to all dealership safety standards.

Skills Required and Physical Demands:

1. Previous parts department experience as stocking clerk, inventory administrator, back counterperson, or parts counterperson.
2. Excellent math, technical, and organizational skills.
3. Ability to lift 70 pounds with a lifting belt (required).

Warranty Administrator

Reports to: Service Manager or Fixed Operation Manager

General Description:
The Warranty Administrator is responsible for the submission, correction, and collection of warranty claims for the maximum amount of credit within the guidelines of the factory

Duties and Responsibilities:

- ❒ Close all warranty claims daily for the accounting office.
- ❒ Submit or "download" warranty claims each night to the factory for reimbursement.
- ❒ Print or save the factory's response for paid and rejected claims daily and keep in a binder or network folder.
- ❒ Apply credits to the warranty schedule.
- ❒ Submit daily, weekly, biweekly, or monthly (depending on the timing of warranty claim credit memos) corrections to the accounting schedule for outstanding and paid warranty claims.
- ❒ Maintain a record (either a log or electronic record) of all claims submitted, date submitted, responses, and payment history.
- ❒ Obtain the service manager's approval for any claims that must be adjusted to policy expense.
- ❒ Monitor the submission of return warranty parts or other required documentation to the factory.
- ❒ Maintain service bulletins and notices that pertain to warranty claims in a special binder.
- ❒ Provide training to the service manager and advisors about warranty coverage and procedures.
- ❒ Assist the cashier or advisors during busy times in the service department.

Skills Required and Physical Demands:

1. Excellent communication and math skills.
2. Ability to lift at least 50 pounds (if the factory requires warranty parts returns).
3. Previous service advisor, cashier, dispatcher, or technician experience.

Service Manager

Reports to: General Manager or Fixed Operations Manager

General Description:
The Service Manager has overall control over customer satisfaction and the profitability of the department.

Duties and Responsibilities:

- ❒ Supervise the service department staff including advisors, technicians, warranty administrators, cashiers, and service receptionists.
- ❒ Manage the sold and unsold hours.
- ❒ Meet with any dissatisfied customers and factory personnel.
- ❒ Maintain high technician productivity with training programs and proper dispatching based on skill levels.
- ❒ Coordinate with the parts manager inventory stocking requirements to obtain a high level of service.
- ❒ Control expenses such as shop supplies and policy adjustments.
- ❒ Increase sales and gross profit by generating new business with mailings, coupons, and customer contact strategies.
- ❒ Control repair order flow and supervise open repair reports and work in process.
- ❒ Maintain conformity with government regulations such as OSHA safety, and hazardous waste disposal.
- ❒ Control the shop tool inventory and shop manuals, purchasing the most current items at the least amount of expense.
- ❒ Promote high customer satisfaction scores and retention percentages.

Skills Required and Physical Demands:

1. Previous service advisor, technician, or related management experience.
2. Ability to communicate with customers, department employees, and other managers with verbal meetings and written reports.

Body Shop Manager

Reports to: General Manager or Parts & Service Director

General Description:
The Body Shop Manager has overall control over customer satisfaction and the profitability of the body shop.

Duties and Responsibilities:

- ❒ Supervise the body shop staff including estimators, painters, and body technicians.
- ❒ Work with insurance adjusters, creating relationships that get the claims paid quickly and according to estimates.
- ❒ Meet with any dissatisfied customers and factory personnel.
- ❒ Maintain high body shop technician and painter productivity with training programs.
- ❒ Coordinate with the parts manager inventory any items on order and any items that need to be returned for credit.
- ❒ Control expenses such as paint and body shop supplies and policy adjustments.
- ❒ Increase sales and gross profit by generating new business with insurance company mailings and advertising.
- ❒ Control repair order flow and supervise open repair reports and work in process.
- ❒ Maintain conformity with government regulations such as OSHA safety, and hazardous waste disposal.
- ❒ Control the shop tool inventory, paint materials inventory, and shop manuals, purchasing the most current items at the least amount of expense.
- ❒ Monitor the maintenance of paint booths, frame straightening equipment, and other large fixed assets to ensure long term usage and value.
- ❒ Control body shop accounts receivables.

Skills Required and Physical Demands:

1. Excellent verbal, math, and leadership skills. These skills may be tested and evaluated.
2. Previous body shop foreman or related management experience.
3. Ability to communicate with customers, department employees, and other managers with verbal meetings and written reports.

Hiring Super Employees

Hiring bad employees increases personnel expenses. A lack of employee productivity costs you overtime. Our industry has never been good at reducing turnover. Hiring and training new employees is expensive. Turnover costs are estimated by experts at 75% of annual salary. This is the cost of training, loss of productivity, and simple things like adding the employee to the payroll system and printing new name tags. In addition to these personnel costs you also have the direct cost of running an advertisement, printing job application forms, performing background and drug tests. In most dealerships prospective employees must be interviewed by two managers. That interview time is wasted time – it doesn't make any money.

Let's start at the top and find the right controller or office manager. The number one reason why I was a successful controller was my office manager. Having a second in command is important if you're going to grow as a controller. If you're a dealer looking for a Super Controller or a Controller/CFO looking for an office manager, here are some ideas to find your next super person and hopefully save some money by avoiding costly mistakes of hiring the wrong person. I find the best controllers or office managers come from within (that's where I came from.) Here are the keys to getting the right person in that position;

- ❒ Quickly identify anyone who has potential on your current staff
- ❒ Hire only those with potential for any new openings
- ❒ Require that each current office manager or controller have a 2nd in command
- ❒ Make sure you know the career path for each of your employees
- ❒ Inform your employees know your plans for their advancement

This process can be used for any position. We've learned from the hiring of technicians that the best come from the apprentice program. Think of your office as a training program for controllers and office managers. Find those who have potential and give them more tasks to do each month. They want to learn more. How do you identify potential and hire only those with potential? You're looking for people who are;

1. Smart
2. Make few mistakes
3. Quick (to learn and do their job)
4. Helpful to others
5. Always asking for more to do or learn
6. Loyal and trustful

Smart is like location in real estate – the three most important things are; smart, smart, smart. Yes, the world needs file clerks too, but not as your super controller or office manager. These two positions must be smart. How to measure smart? Ask to look at high school and college transcripts. You might want to get legal advice about requesting transcripts – since it indicates age, so instead put the GPA on your application with a note that it will be verified before employment (you can require the transcript after the job offer to verify.) I included my own transcripts with any job that I applied for and I think it helped me get hired. What if you don't have anyone with potential and need to hire someone with potential? Here's a perfect recruiting advertisement – you can run in two areas –

Accounting Manager Trainee and Manager Trainee
Manager Trainee (Also run as "Accounting Manager Trainee")

We are looking for a management trainee for our accounting department (or dealership.) No experience is required, but you must like supervising an accounting staff and managing a computer system. Accounting (or Automotive/Business) experience or (a business-focused) education is helpful but not required if you like working with numbers (people,) learning new things, and working with other managers/computers.

What do you do with this person? Are you kidding? Every accounting office/dealership that I visit is buried. There is always work to be done. Create a self-study training program that includes online classes for Word, Excel, your DMS system, your manufacturer's business management guide, etc. This should take a year and you need to spend serious time with this person teaching them the business. By hiring from within or getting a lower cost manager trainee, you'll avoid getting a high priced person from the outside that might struggle adapting the many "quirks" of our industry.

Every CFO needs a controller that is next in line to be the next CFO. Every controller needs either an office manager or assistant controller. Titles are important to your staff and they don't cost you anything. Another thing that doesn't cost you anything is information gathering. Ask a cashier during a review "what other position in the dealership would you enjoy doing?" If you cashier responds with service advisor or greeter, then you don't want to spend a lot of time training her on how to clean the sublet schedule. Instead, let the service manager know so he or she can expand the cashier's duties like setting future appointments, making follow up calls and performing service campaigns. If instead your cashier wants to be a payroll clerk or wants your job, then start giving her schedules to clean, accounts to reconcile, or show her how to break out the service department expenses for the manager each month. After you have that conversation, make sure your cashier knows that you will do your best to move her to a new position as openings come up. Often an employee will stay with you longer – even if they don't like the job, if they know that in six months, they will get moved into the office or out to be a greeter.

Performing Employee Reviews

Now that you have hired, staffed and trained your department, how to you keep employees from leaving? Both reviews and bonuses are a way to not only let your employee know that you appreciate them, but since we are a profit generating industry – they have the added benefit of making more money. How can an annual review help you make money? It is easy – tell your employees what you need from them and it is human nature to please. I know this is hard to believe as you look out on your sales force or staff – but it is true. Nobody goes intentionally into work today with the objective to do a bad job. Employees do not like making mistakes and certainly do not enjoy you pointing them out. How do mistakes happen? The biggest reason is that employees didn't realize that it was important to careful. A review is the perfect time to go over these issues and put the greatest emphasis on those items that matter the most for that position. Let's take two examples; a back counter parts employee and a service cashier. In the Free downloads at SandraJerome.com is a spreadsheet with this formula. Looking at the cashier's tab you can see that there are quite a few areas for improvement.

Employee Review Form		
Employee Name: Sue Smith		**Position: Cashier**
	Score	**Weight**
Attendance	8	9
Accurate reconciliation of cash	2	10
Friendliness to customers	7	9
Limited overtime	2	6
Conformity to corporate rules	5	7
Clean work area	9	3
Weighted Review Score	5.20	
Goal Score	7.00	

In this example this cashier achieves one of my greatest expectations – she shows up every day. This makes a nice start to the review process and I recommend that you try to start with a positive item. Here's how this review would sound, "Sue, I'd like to show you the new method we will be using here at Sandy Chevrolet to perform your annual review. As you can see, you almost have perfect attendance and we consider a high expectation of your position – it rates a 9 on a scale of 10. Another high expectation is that the daily cash reconciliation equal zero. Unfortunately based on the input from the office manager, you score very low. She has provided me with this report of a random two week period. You only had one day where the cash reconciled to zero. Most of the days where it did not, you not only were considerably off, but it took you hours to come to that amount. I feel that this issue has directly contributed to your low score on overtime. What I would like for you to do is go back to your desk and list out some of the reasons why you are having difficulty reconciling cash sales and please list anything you think we can do to help you. Based on this review, I cannot recommend a raise at this point, but if you can bring this score up to a 6 in the next month, then I'll recommend a small raise and if you can reach the goal of a 7, then it would be a much larger raise. I'd also like to compliment you on your friendliness to customers, but you'd have a higher score there if you would extend that friendliness to the other employees especially the service advisors. For our office staff, the other employees are considered our customers. You have a neat work area and you scored very high, but we do have some issues with your attire and I'm going to circle the page in our company handbook where these issues are addressed."

See how easy that is? Your objective is for that cashier to do a better job of controlling the cash drawer and reduce her overtime. Improvements in both these areas mean more profit, so this review can make you money. Let's look at the parts counterperson's review.

Employee Review Form		
Employee Name: Billy Ray	**Position: Parts Back Counter**	
	Score	**Weight**
Attendance	9	9
Fill rate	4	10
Cooperation with technicians	3	10
Limited overtime	2	8
Conformity to corporate rules	9	7
Clean work area	7	5
Weighted Review Score	5.41	
Goal Score	7.00	

Again, you would complement him on his attendance and conformity to company rules but point out the low fill rate and cooperation with technicians. If this back counterperson were to hit the goal score by improving your fill rate and increasing technician productivity –that would mean more profit. He also needs to get his overtime under control and that reduces expenses.

I recommend using this spreadsheet and showing the employee what they need to do to hit the goals. You can make it more specific if you want by providing a raise percentage for each goal step. Doing a review is uncomfortable for all participants, so if you use something like this spreadsheet – you can stay focused on your objectives and their goals instead of turning the session into a painful discussion. Write down bullet points to help you through the review and provide these bullet points along with the scoring page to the employee. I'd leave room for them to enter comments. The form on the following page enables you to convert this spreadsheet into something that can be stored in their file and used during the review session.

Sandy Chevrolet

Employee Review Form

Employee Name:____________________Date:____________

Reviewed by:_______________________________________

Scoring - 10 is excellent (no room for improvement) 5 is good 3 is fair 1 is poor

Basics	Score	Profit	Score	Customers	Score
Performs all job duties		Suggests ways to save money or increase profit		Treats other employees with respect	
Attention to detail		Reduces paperwork and waste		Helps profit generating employees do their jobs better	
Sense of urgency		Uses spare time to organize and clean		Courteous and friendly to customers	
Accurate and tries to learn more		Helps other staff get their job done		Pleasant phone skills	
Conforms for rules and regulations		Avoids overtime		Makes management aware of any customer complaints	

Current Score____ Prior Score ____ Previous Review Date_________
Last wage increase date _________ Last wage increase $______
Recommended increase by reviewer $_______
Amount approved by supervisor/dealer $_____ Signature:_________________

Accomplishments: (can also be completed by employee)

____

____

____

Training recommended:

____

____

Areas that could use improvement:

____

____

Comments:

____

____

Career path:

For office use only

Dealer approved ___
Updated in payroll___
Next review date:_________
Flagged in Outlook:___
Notes:

Employee Pay Plans

The important thing to remember when analyzing pay plans is to determine if the pay plan is within expense guidelines and the goals of the dealership. If it is the dealership's goal to provide good customer service, then this needs to be in the pay plan.

Dealership pay consist of 3 items

1. Salary
2. Commission
3. Bonuses

The salary, base, or guarantee helps you keep good people in bad times. Another element is commission. This is a performance reward base on sales, gross profit or other measurable amount. The final element is a bonus if the employee reaches *YOUR* objectives. That's the most important element of a pay plan. If a pay plan doesn't get the results that you wanted from that department, then you need to change it. For each employee below there is an example that puts an emphasis on each element and a range of amounts. These are only guidelines since salary and bonuses vary greatly from region to region and type of product. The key to compare what the employee is paid on – for example you wouldn't normally pay a Sales Manager on total dealership net profit since that person has little control over the bottom line and fixed expenses like rent and insurance. An employee needs to be rewarded for the items they can control and make a difference in improving.

What percentage should an employee be paid salary vs. incentive? An employee that is highly productive and generating gross profit would have 70 – 100% of their pay as incentives (bonus plus commissions. Technicians and sales people fall into this category. On the other end of the spectrum are office managers or controllers who would have 90 – 100% of their pay based on salary. If you are trying to reach certain objectives, then the person who can get you those results would have a high bonus. For example, a Customer Service Manager might have 50% salary and 50% bonus based on Customer Satisfaction Scores. The following pages have examples of the major positions at a dealership. Within each position, some of the example put a higher emphasis on commissions or bonuses and others have a high salary or base. In order to get a pay plan accepted, you might need to offer a guarantee for a certain amount of months.

General Sales Manager

The General Sales Manager's pay is normally similar to the General Manager except vehicle gross instead of full dealership gross profit. Like the General Manager, the amount of commission and what fund depends on how much they control adjusted selling gross and vehicle net profit.

Example 1	High	Low		High	Low
Salary or base				$ 2,500.00	$ 1,000.00
Commission					
Total Vehicle Gross	$310,936.00	$187,707.00	2%	$ 6,218.72	$ 3,754.14
Bonus items					
CSI/QCP scores				$ 1,000.00	$ 250.00
Increased volume				$ 1,200.00	$ 800.00
Total Pay				$10,918.72	$ 5,804.14

Example 2	High	Low		High	Low
Salary or base				$ 7,500.00	$ 5,500.00
Commission					
Total Vehicle Selling Gross	$ 82,692.32	$ 48,838.00	3%	$ 2,480.77	$ 1,465.14
Bonus items					
Advertising per unit less than $200				$ 750.00	$ 200.00
No unit over 60 days				$ 200.00	$ 50.00
Total Pay				$10,930.77	$ 7,215.14

Example 3	High	Low		High	Low
Salary or base				$10,000.00	$ 8,000.00
Commission					
Total Vehicle Net Profit	$ 36,619.94	$ (9,400.00)	10%	$ 3,661.99	
Bonus items					
Gross Profit per unit				$ 750.00	$ 500.00
Sales per Salesperson				$ 500.00	$ 300.00
Total Pay				$14,911.99	$ 8,800.00

Sales Managers

The sales manager pay is almost identical to the General Sales Manager, other than a greater emphasis on commission based on gross profit less controllable expenses (Example 2) vs. salary.

Example 1	High	Low		High	Low
Salary or base				$ 1,500.00	$ 1,000.00
Commission					
Total Vehicle Gross	$310,936.00	$187,707.00	2.5%	$ 7,773.40	$ 4,692.68
Bonus items					
CSI/QCP scores				$ 700.00	$ 250.00
Increased volume				$ 500.00	$ 150.00
Total Pay				$10,473.40	$ 6,092.68

Example 2	High	Low		High	Low
Salary or base				$ 5,000.00	$ 2,500.00
Commission					
Total Vehicle Selling Gross	$ 82,692.32	$ 48,838.00	4.0%	$ 3,307.69	$ 1,953.52
Bonus items					
Advertising per unit less than $200				$ 750.00	$ 200.00
No unit over 60 days				$ 200.00	$ 150.00
Total Pay				$ 9,257.69	$ 4,803.52

Example 3	High	Low		High	Low
Salary or base				$ 5,000.00	$ 5,000.00
Commission					
Total Vehicle Net Profit	$ 36,619.94	$ (9,400.00)	15.0%	$ 5,492.99	
Bonus items					
Gross Profit per unit				$ 500.00	$ 250.00
Sales per Salesperson				$ 500.00	$ 300.00
Total Pay				$11,492.99	$ 5,550.00

Variations;

- ❒ Rate of gross profit is a step percentage based on
  - Average gross profit
  - Units sold
  - Age of inventory
  - Zone penetration (new)
- ❒ Type of Gross Profit
  - 2% of own department (used) and 1% of new
  - 3% of own department and 2% of F&I

F&I Manager

The F&I Manager can be either a product salesperson where there is a greater emphasis on commissions vs. salary, or more of a department manager, where the salary is higher. For each of the examples there can also be a variation in the percentages – for example

Reserve commission	1%
Aftermarket	10%
Service contracts	5%
Credit Life and AH	7%

The percentage is based on how much of the product is retained. Aftermarket products have no chargebacks so dealerships retain all the income. Reserves have high chargebacks as customer pay off their vehicles early.

It is also the custom to either distribute a portion of the total chargebacks to each F&I manager – or subtract only their own chargeback from their commissions.

Example 1	High	Low		High	Low
Salary or base				$ 750.00	$ 500.00
Commission					
Total F&I Products for department	$ 86,197.00	$ 15,507.00	7.0%	$ 6,033.79	$ 1,085.49
Bonus items					
CSI/QCP scores				$ 250.00	$ 150.00
Penetration %				$ 500.00	$ 200.00
Total Pay				$ 7,533.79	$ 1,935.49

Example 2	High	Low		High	Low
Salary or base				$ 1,500.00	$ 1,200.00
Commission					
Total F&I Product sold by manager	$ 82,692.32	$ 48,838.00	12.0%	$ 9,923.08	$ 5,860.56
Bonus items					
F&I income per unit				$ 500.00	$ 150.00
Total Pay				$ 11,923.08	$ 7,210.56

Example 3	High	Low		High	Low
Salary or base				$ 2,000.00	$ 2,000.00
Commission					
Total Net F&I Income less C/B	$ 36,619.94	$ 23,425.00	15.0%	$ 5,492.99	$ 3,513.75
Bonus items					
Vehicle receivables age				$ 500.00	$ 250.00
Total Pay				$ 7,992.99	$ 5,763.75

Salesperson

Sales pay plans have various rates from 20% to 35% and this percent is often determined by the number of units sold. Bonuses (spiffs) are used to reach various management objectives like selling an older unit, or providing good customer satisfaction. It is rare for a salesperson to get a salary or base unless it is a "one price" dealership or the salesperson as show in Example 3.

Example 1	High	Low		High	Low
Salary or base				$ -	$ -
Commission					
Vehicle Gross	$ 15,325.00	$ 11,252.00	30.0%	$ 4,597.50	$ 3,375.60
Bonus items					
Spiffs				$ 250.00	$ 150.00
Total Pay				$ 4,847.50	$ 3,525.60

Example 2	High	Low		High	Low
Salary or base					
Commission					
Vehicle Gross	$ 19,525.00	$ 9,525.00	25.0%	$ 4,881.25	$ 2,381.25
Bonus items					
F&I income	$ 2,342.00	$ 1,252.00	5.0%	$ 117.10	$ 62.60
Total Pay				$ 4,998.35	$ 2,443.85

Example 3	High	Low		High	Low
Salary or base				$ 1,500.00	$ 750.00
Commission					
One price per unit	15.00	10.00	$150.00	$ 2,250.00	$ 1,500.00
Bonus items					
Unit bonus				$ 500.00	$ 250.00
CSI bonus				$ 200.00	$ 150.00
Total Pay				$ 4,450.00	$ 2,650.00

Other Salesperson Incentives;

- End of the year bonus; $5 per vehicle into a fund (former Christmas club bonus)
- Top salesperson bonus (based on gross, units, CSI and other items) $500
- Gross Profit Step Bonus (after packs)
  - Over $15,000 $200
  - Over $20,000 $500
  - Over $25,000 $700

Internet or BDC Manager

The above Example 3 is very common for an Internet or BDC manager with the per unit price being everything from $10 to $100, depending on the involvement of the person. If the Internet Manager or BDC manager only makes an appointment, then it is closer to $10. If they make the appointment, greet the customer and do most of the deal processing, then $100. If they are paid from Example 1 and share the deal, the rate is 5% - 15%. If they are paid off Example 1 and have the whole deal, the rate is 15% - 35%.

Customer Relation Manager

- Bonus on CSI
- Survey call volume

Service Manager

Most service managers are paid a salary and bonuses on high technician productivity, controlling expenses such as shop supplies and policy adjustments, increased gross profit, and promoting high customer satisfaction scores and retention percentages.

Example 1	High	Low		High	Low
Salary or base				$ 3,500.00	$ 2,500.00
Commission					
Total Service GP	$ 92,393.00	$ 73,234.00	5.0%	$ 4,619.65	$ 3,661.70
Bonus items					
CSI/QCP				$700.00	$500.00
Fixed Right First Time				$400.00	$300.00
Total Pay				$ 9,219.65	$ 6,961.70

Example 2	High	Low		High	Low
Salary or base				$ 2,500.00	$ 1,500.00
Commission					
Total Service Selling Gross	$31,323	$12,534	10.0%	$ 3,132.30	$ 1,253.40
Bonus items					
Tech Productivity				$400.00	$300.00
Hr per RO				$200.00	$100.00
Total Pay				$ 6,232.30	$ 3,153.40

Example 3	High	Low		High	Low
Salary or base				$ 6,000.00	$ 4,000.00
Commission					
Total Service Net Profit	$4,236	($544)	15.0%	$ 635.40	
Bonus items					
Warranty AR				$700.00	$300.00
Comebacks/Goodwill under budget				$300.00	$100.00
Total Pay				$ 7,635.40	$ 4,400.00

Assistant Service Manager

This position is either paid similar to the service manager, a flat salary plus some of the same bonuses as the above examples, or if the Assistant Service Manager was a former Shop Foreman, then a small amount per tech hour paid plus bonuses (see the shop foreman examples.)

Body Shop Manager

Same as service manager, but often a bonus is paid on controlling body shop supplies (a percentage of excess.)

Service Advisor

It is the goal of the service advisor to have a high labor and parts sales per repair order and high customer satisfaction scores. Most service advisors are paid on a percentage of labor and parts sales.

Example 1	High	Low		High	Low
Salary or base				$ 2,000.00	$ 1,500.00
Commission					
Total Labor and Parts Sales	$ 35,125.00	$ 18,252.00	3.0%	$ 1,053.75	$ 547.56
Bonus items					
CSI/QCP				$500.00	$250.00
Total Pay				$ 3,553.75	$ 2,297.56

Example 2	High	Low		High	Low
Salary or base				$ 2,500.00	$ 1,500.00
Commission					
Total Labor and Parts Gross profit	$27,585	$15,252	7.0%	$ 1,930.95	$ 1,067.64
Bonus items					
Hr per RO				$500.00	$400.00
Total Pay				$ 4,930.95	$ 2,967.64

Example 3	High	Low		High	Low
Salary or base				$ 4,000.00	$ 3,500.00
Commission					
Total Labor Sales (whole department)	$153,252	$92,858	1.5%	$ 2,298.78	$ 1,392.87
Bonus items					
Comebacks/Goodwill under budget				$300.00	$100.00
Total Pay				$ 6,598.78	$ 4,992.87

Variations;

- ❒ Labor type determines rate- Customer pay (10%, Warranty 5%, Internal 2%)

Technician

- Per flat rate hour
- Bonuses based on
  - Fixed right the first time
  - Low comebacks
  - Low shop supplies
  - Ready when promised
  - Productivity, Efficiency, and Opportunity percentage

Dispatcher and Shop Foreman

Pay is either tied to Technician pay;

- .50 - $1.00 of hour flagged by the shop
- Fix it right bonuses
- Team bonuses

Or tied to Advisor pay;

- 2% of labor gross profit
- 1% of labor sales
- Ready when promised
- Under budget on comebacks/goodwill

Booker, Biller, or Cashier

- Paid a bonus if they find parts not on repair order for a labor operation code – 10%
- CSI
- Percentage of shop supply fee collected
- $1.00 for each email address received from the custom

Warranty Administrator

- Percentage of all claims paid 1-2%
- Bonus for amount current; $500 if over 90%
- Step percentage based on amount current
  - 90% 2% of claims
  - 85% 1.5%
  - 80% 1.0
- CSI bonus

Parts Manager

Pay is based normally on gross or net profit and bonuses for internal controls to prevent shortages and obsolescence, generate gross profit, and control expenses.

Example 1	High	Low		High	Low
Salary or base				$ 2,500.00	$ 2,000.00
Commission					
Total Parts GP	$ 152,525.00	$ 43,653.00	1.0%	$ 1,525.25	$ 436.53
Bonus items					
Obsolescence less than	5%	10%		$600.00	$500.00
Total Pay				$ 4,625.25	$ 2,936.53

Example 2	High	Low		High	Low
Salary or base				$ 4,000.00	$ 3,500.00
Commission					
Total Parts Selling Gross	$73,252	$35,325	5.0%	$ 3,662.60	$ 1,766.25
Bonus items					
Inventory turns	7 times	6 times		$500.00	$400.00
Total Pay				$ 8,162.60	$ 5,666.25

Example 3	High	Low		High	Low
Salary or base				$ 2,500.00	$ 1,500.00
Commission					
Total Net Profit	$42,522	$23,252	10.0%	$ 4,252.20	$ 2,325.20
Bonus items					
CSI bonus				$700.00	$500.00
Total Pay				$ 7,452.20	$ 4,325.20

Other pay options;

- Inventory age
  - 90% current $500
  - 85% current $300
  - 80% current $200
- Bonus for sales per employee
  - $20,000 $100
  - $22,000 $200
  - $25000 $500
- Different percentages for sale type
  - Repair order gross 1%
  - Counter retail 5%
  - Wholesale gross 7%

Parts Counter

- Wholesale salespeople, 1 – 2% of sales
- Back counter 2-3% of repair order sales
- Front counter 5-10% of retail gross profit

General Manager

There are three levels of profit to pay a General Manager on; Gross Profit, Adjusted Selling Gross and Net profit. If a General Manager has control over most expenses

– over 75%, then net profit is the best option (Example 3.) If on the other hand, the General Manager only controls personnel, advertising, floor plan, and other "selling gross" expenses, then the selling gross concept works best. What should you include in selling gross? It would be your gross profit less those expenses the General Manager can control. If your General Manager is more like a General Sales Manager and the dealer controls most expenses, then pay on the Gross Profit as in Example 1. The bonus items round out the pay in months where the gross or net is low, but objectives were met.

Example 1	High	Low		High	Low
Salary or base				$ 5,000.00	$ 2,500.00
Commission					
Total Dealership Gross Profit	$ 505,936.00	$ 343,937.00	1%	$ 5,059.36	$ 3,439.37
Bonus items					
CSI/QCP scores				$ 1,000.00	$ 50.00
Total units sold				$ 800.00	$ 200.00
Total Pay				$ 11,859.36	$ 6,189.37

Example 2	High	Low		High	Low
Salary or base				$ 7,500.00	$ 3,500.00
Commission					
Total Dealership Selling Gross	$ 164,751.32	$ 98,760.00	2%	$ 3,295.03	$ 1,975.20
Bonus items					
Advertising per unit less than $200				$ 750.00	$ 200.00
Sales per employee				$ 200.00	$ 50.00
Total Pay				$ 11,745.03	$ 5,725.20

Example 3	High	Low		High	Low
Salary or base				$ 10,000.00	$ 5,000.00
Commission					
Total Dealership Net Profit	$ 111,406.94	$ 12,324.00	5%	$ 5,570.35	$ 616.20
Bonus items					
Age of vehicle inventory				$ 500.00	$ 200.00
Expense reduction				$ 500.00	$ 150.00
Total Pay				$ 16,570.35	$ 5,966.20

CFO, Controller and Office Manager

The objective of the CFO, Controller, or Office Manager is to provide a smooth flow of paperwork through the office, reduce expenses, and protect the assets of the company. The following ideas are incentives to help reach these goals for this position and recently bonuses have been added for the whole office staff.

- 1% - 5% of net profit
- 10% of any savings ideas
- Bonus for asset management
    - Contracts in transit, vehicle receivables 90% current, $200
    - AR 80% current $200
    - Warranty claims 80% $200
    - 10% of amount transferred into Cash management fund
- Financial statement
    - Done by 10th $100
    - By 5th $200
    - By 3rd $500

Car Billers

- $1 for each deal billed
- Contracts in transit, vehicle receivables 90% current, $200
- Bonus for days between delivery and finalized in accounting

AR Clerk

- AR 80% current $200
- AR 90% current $500

Office staff

- Financial statement done by 3rd working day, 4 hours free time off
- Small percentage of net profit
- Flat amount for emails collected
- Cashiers – percent of parts found missing on repair orders or for having no open completed internals at the end of the day

Budgeting for Compensation

These metrics are only a guideline but can be a starting point for determining if you are close to other dealerships.

- ❒ Total new personnel expense should be 30%.
- ❒ Used should be 25%.
- ❒ The average compensation for a service manager is 5-10% of the service gross.
- ❒ Total service personnel expense should not exceed 35% of gross.
- ❒ The average compensation for a parts manager is 8% of the parts gross.
- ❒ Total parts personnel expense 40%.
- ❒ Body shop personnel expense should not exceed 50%.

Files in the Contact – Free Stuff at www.SandraJerome.com

Expense Metrics

Super Budget Worksheet

Expense Attack Worksheet

Employee Review Form

Advertising Budget vs. Actual

Job Functions Worksheet

Sales Commission Analysis

Vacation Accrual Spreadsheet

Employee Weighted Review

All the Job Descriptions in Word Format

Super Controller Book 3 - *Time and Tasks*

Introduction to Time and Tasks

Do you need more time in the day? Are you getting buried by too many tasks to complete each month? Super Controller 3 is your guide to Time & Tasks. The goal of this guide is to help you get more time in the day and in return enable you to become more of a Super Controller. I wrote the first **Super Controller Guide** to introduce office managers and controllers to the rest of the dealership. Hopefully you've read this guide and have learned how each of the other departments function and now how you can help others do their jobs better. The key to being a Super Controller is being in control of the whole dealership instead of just the accounting the office. Next, I **wrote the Super Controller-2 Expense and Employee** that takes the controller back into the office to tackle two problem areas; Expense Control and Employees but with an objective to reduce the expenses of the whole dealerships. Over the past years, when I would give a seminar on **Super Controller I and 2**, a common complaint was that most controllers lack the time to tackle the great ideas in these guides – so I went back to the beginning and starting writing this guide as a way to help you get the time you need to execute the ideas in the first two guides.

In the **Time** section you'll learn how to;

- ❐ Be in better control of your time
- ❐ Find things that you'd do with your extra time
- ❐ Eliminate time wasters
- ❐ Make an action plan to get more time

In the **Task** section you'll learn how to;

- ❐ Become a task-based Accounting office
- ❐ Determine if a Task should be done daily, weekly, monthly, during the close, or yearly
- ❐ Use skill levels in delegating tasks
- ❐ How to make Tasks easier to delegate and train
- ❐ Over 300 Tasks with Tips for you to use to develop your "How to" training guide

As always, my “take” on these two topics is a little different as most controllers, office managers, CPAs, or consultants. Hopefully you’ll get some ideas to help you in these two areas – they were always a challenge for me, but over the years of being a controller and a consultant, I discovered some ways to make the job easier and get better results. Read on!

Understanding Time in a Dealership

I visited a dealership that had a controller that that seemed to work bell to bell, Saturday, - sometime Sunday (when he said it was nice and quiet) and INSTEAD of the dealer appreciating his dedication which I had pointed out, he said that the controller wasn't very organized and had to work longer to do the same stuff as his prior controller – weird.

"People don't judge you on how many hours you put in;
Rather on of the VALUE of the time you work."

What do you do All Day?

Your first homework is to for one day, write down everything you do all day. Just use a simple yellow legal pad and make notes as the day goes by. Include in this list this anyone who comes into your office, phones, texts, or emails you and why. It is a big chore but as you review this the next day, I think you're going to be surprised.

Take the list and write next to it with a red pen three categories;

T – Tasks that you do as part of your job; reconciling accounting, paying the sales tax, checking the bank balance. You can go further and indicate TD as Task Daily, TW as Task Weekly, etc.

F – Fire drills. I call this any unplanned issue; a customer who lost a check, an employee upset about their pay, someone wrecked their demo or got hurt in the shop, and the dealer with about any reason…

M – Monkey. This is anyone who came into your office, emailed, texted, phoned to get something, ask you a question, tell you a story, or just chat with you about their weekend. See the Monkey section to understand better. Normally anything that is not a T or F is a Monkey.

What do you WANT to do All Day - Goals

As a controller, I discovered that I spent about 2-4 hours a day doing my tasks. I was obsessive about trying to get my daily tasks done by 10AM so I could get my monthly tasks that I was working on completed by the 5th. Then I'd have lots of free time in the day and month. What could you do with this free time? Here are some of the things that I did during my 25 years.

- Learn your DMS better. I ended up going through about 10 conversions. I was always willing to learn a new DMS system, in fact I went after work to help any of my friends that were getting a new DMS to help them do the setups. Whenever we got a new module, I'd learn it and help install. My most recent at the dealership was when they wanted to go on automated dispatching. I did all the training on it (my service manager was way too busy) and got it working properly. Same with the new CRM software. It is how I became an expert at DMS systems and eventually designed my system, DealerStar.

- Get involved in HR. I don't like anything to do with employees, but I know of other controllers that have become experts in HR and payroll. One even went on to become the head of HR for a huge school district with a great salary and benefits. It is the one area in our business where we do things the same as other companies.
- Profit Projects. At my most recent dealerships, my bonus was a 10% of any profit/savings plan I could come up with, so I spent a lot of the day looking for a way to make more money. Much of this work I documented in Super Controller II – Expenses and Employees.
- Do Yoga/take walks. I never seemed to have time to do this at home, so I'd put my walking time in the middle of the day and walk around the dealership, or go into the break room and do Yoga.
- College, CPA. I completed college and passed my CPA exam while working dealerships (it took 7 years!) When I had free time, I'd do my homework using my current set of books (cash flow statements, etc.) and many of these spreadsheets I modified for homework ended up being part of this book.
- Train for another job; F&I manager, service advisor, General Manager. At one point in my career I thought I'd like to be a dealer/General Manager. My current GM didn't like to go to any meetings or training, so I'd take all of those for him and started getting more involved in the other portions of the dealership. Much of the information in the original Super Controller I learned as I was an Assistant GM and taking these courses. I know of another controller that decided she wanted to learn how to be an F&I manager and another that trained to be an service advisor.
- Excel classes. There are great online classes or you can take one during the day at a local community college. I became an Excel guru that helped me do my job faster and then ended up writing Super Excel with Jack Ross (who is ***THE*** Excel guru of all times!)

If all of this sounds attractive to you, it is important that your list is mostly "T"s and you minimize the F and M on your list. Let's talk first about those Monkeys!

How to get the Monkey off your Back

I was once working at a dealership, not with the controller this time, but rather deeply involved in their computer database; they were about to convert systems, so I was running report to back it up. I was working at a terminal right outside the controller's door. As I'd wait for various jobs to run, I got lots of time to study this amazing Super Controller. She had that multi-tasking thing down – but I was amazed at the amount of traffic in and out of her office. This was at a time when she was supposed to be learning the new computer system and doing all those important setups that I was there helping her with – but instead people came in and out. So I started making a list. By the end of the day, 47 people had entered her office. Many didn't think they had interrupted her, but most people muttered while they went in and out, shuffled through the stuff on her desk or got something out of a file cabinet and each time, she'd pause and look up. And later we wondered why all the sublet sales in the new computer went to used wholesale. She was horribly distracted during the day. There is a book, **The One Minute Manager Meet the Monkey** and the writer discusses how everyone who enters your office has a monkey on their back and their objective is to get the monkey to jump from their back onto your back. I find most controllers and office managers have a whole gang of monkeys on their back by the end of the day. That end up exhausted and tired from carrying the load. Who is the worse Monkey? It is not the dealer or GM as you might think. It is you own office staff. They come into your office all day long! The purpose of this book is to find ways to keep them busier (they really have more time that you think) and out of your office by enabling them to do their tasks without disturbing you. The next time someone comes into your office – listen carefully. Are they trying to unload their Monkey?

"I can't balance cash sales"
"I can't find the Jones deal"
"We're out of coffee in the lounge"
"The bank just called and needs the …."
"Have you seen my stapler?"
"Sue just said, blah, blah blah…."

One of the best ideas is to remove the chairs from your office. This keeps the drama queens from coming in and dumping their woes or telling you about their crazy weekend.

File Cabinets – Perfect Placement

Nobody likes to file and nothing is more wasteful than a big row of file cabinets against a wall. Try to disperse the file cabinets throughout the office. You'll see in the filing section of tasks, that I prefer 4 drawer and 2 drawer cabinets be placed next to desks. For example AP files are next to the AP desk. Car deals are next to the biller desks. Warranty books, extra sets of keys – keep them in parts. That is who will have to order one if they are missing. ***No controllers/office managers should have a lot of file cabinets in their office. Place them outside your office, especially if others must get items from the cabinet.***

Managing your In Box

Remove your In Box from your desk and put it on top of the file cabinet you just placed outside your office. You can still see it, but it enables people to give you things without coming into your office and disturbing you.

Your inbox in Outlook or even on your desk should only contain 4-5 items. Make sure as new emails or tasks arrive on your desk that you either delegate to your staff or assign to your daily, weekly, or month tasks.

Signing Checks, Contracts

Don't sign checks and contracts throughout the day, instead make a time when you'll sign them and be faithful to that time. 3pm is a good time and put the bucket outside your office on the file cabinet. Sign them and put them back there by 4pm.

Phone Calls

Let everyone know that they can email you – or employees can instant message you. If you switch to this method – be good about checking your email every half hour. Put on a good filtering system so you don't have to read junk mail. By the way, I only communicate via email!

Opening the Mail

Do you think that the new head of 3M opens the mail? In most companies, opening and distributing the mail is done by the lowest paid employees; the mail room. Consider having your receptionist open the mail and hand it out. Do you need to monitor checks received or complaint letters? Fine, create a report for her to complete. Want to make sure nothing odd is happening? Great, then look through the mail once a month – or once a week just to keep everyone on their toes.

Computer problems

You pay your DMS company thousands per month to support your employees. IT is often more than you make per month! Make a little note of how to call support (the phone number, your client number, etc.) and put it on each terminal. Let employees know that they are to call support unless it's an emergency like a customer waiting or system down.

Meetings

My favorite poster from www.despair.com is "Meetings - None of us is as dumb as all of us together." Most managers meetings turn into bitch sessions and waste the time of your managers. If you want to see this, just glance over and see how bored your body shop manager is. Why not have focused meetings like AR collection meetings, prospect status sales meetings, heat sheet meetings for contracts in transit? These should be short and focused meetings and find the smallest place to hold them – like the sales office where everyone has to stand up. Have agendas and stick to them. Big comfortable meeting rooms are time wasters.

Petty cash/spiff checks

You are mostly controlling a mulit-million dollar company. Why then try to control a two hundred dollar fund? There are much bigger balances that need your attention. Give this task to your payroll clerk or another trustworthy staff person.

Quick Close in 3 Days or Less

I was a terrible controller when I was closing the books. I'd put a big sign on my door and keep it tightly closed. It was an instant tongue-lashing to anyone who disturbed me during the dreaded "close!"

As I grew into a Super Controller, I realized that I didn't like myself during those days and the only way to make it better was to simplify the process and make it go quicker. This is a revised Quick Close updated from my original Super Controller Guide, but still in 3 Days that will hopefully help you be a nicer person and get the books closed quicker. The quicker you get them closed, the sooner you can get back to your normal daily routine and be a more likeable person.

Before the last day:
___Purify all schedules. Working on schedules should NOT be a part of the close – this includes the vehicle schedules. *As a suggestion, use the following guideline for purification*
Vehicle Receivables/Contracts in Transit – daily.
Vehicle Inventory – weekly and create overage unit lists.
Floorplan – monthly, after the close when you receive your floorplan statement and weekly for small errors.

Accounts Receivable *– weekly for small errors and again before sending out the statements on the 25th.*
Accounts Payable *– weekly looking for debits and after the statements have been paid on the 10th and the 20th, purify.*
Sublet Repairs *- weekly.*
Warranty Claims *- purify after each credit memo is applied (weekly) and keep the final purification after last credit memo is processed for the month.*
___Make all Standard Entries
___Accrue advertising based on budget.

Last Day – Work in Process Day

___Look at a list open repair orders/tickets. Close and update to accounting any that you can.
___Verify that cash sales is zero.
___Make sure all Internals have been closed for vehicles.
___Enter all AP invoices for vehicles (to make sure the cost is correct when they post today.)

Day 1 – Getting the stuff into the office
___Cut off Car Deals at noon.
___Post Final Payroll or Accrue
If you pay through the 31st, then complete the end of month payroll and post, otherwise accrue. *Sales commissions should always be accrued on the deals and do not affect the bottom line except for payroll taxes. Estimate and accrue.*
___Post all car deals by the end of the day.
This is a tough step. It requires that someone help the biller or billers to post the deals into accounting; possibly the office manager or the controller. Nobody leaves until all the deals are posted. A good reason to become a task-based off and "flag" car deals.
___Adjust Work in Process if necessary.

Day 2 – Making sure it is all "In" and getting the "number

___Reconcile vehicle deals posted and grosses with the deals per F&I report. Give report to Dealer/GM
___Obtain the flooring interest charge, bank charges for the month and accrue.
___Enter the remaining Accounts Payable invoices.
The first ones to be processed should be expense payables. Do not allow any invoices to be posted in the afternoon to the vehicle inventory accounts. It is not necessary to post the factory parts statement - just the invoices. This is just another vender and it should be accrued prior to the closed-not reconciled.

___Post the remaining general journal vouchers for accruals.
___Run a DOC and accrue bonuses.
The DOC should have every line item that they need to accrue the bonuses and have an final income statement that reflects the data in the Operating Report. Compare the DOC to last month's - or run a comparative to check that all expenses are accrued and everything looks reasonable. If they currently do not have a functional DOC take the time to create a useful DOC as this will really help they with accruing bonuses and doing comparative analysis.
___Post final bonus general journal vouchers.
___Run the final DOC and leave with dealer/GM and ask for approval to close.
___At night print/run mini-schedules and reports that they need for additional financial statement information such as aging amount or memo counts.

Day 3 – Financial Statement
___Obtain from payroll personnel counts for every department. *You may want to give the payroll clerk a copy of the personnel count section of the financial statement and request that it be completed*
___Make Memo entries and print the Financial Statement.
___Print the financial statement on blank paper and photocopy onto overlays. It is really unnecessary to provide original copies to anyone. If they have laser printers consider the new financial statement print software available from the computer companies.
___Make up backups and print necessary "hard copy" information at night.

Four Hours Free Time Off - Management's Agreement

For a 3-day Close

Let's close the books in 3 days! Why? It enables everyone to get into the current month. Managers concentrate on the next month's business, your staff is ready to help customers and the office is more helpful if that terrible close is done and behind us. You'll impress your 20 group, banks, factory – everyone! We'll be more of a daily and weekly reporting machine instead of one that is only focused on monthly results – weeks later.

What does it take?

- ☐ First of all, you agree to give the office staff 4 free hours off during the next month for free – including the Controller and office manager if they get it done.
- ☐ You will enforce the cutoff of the deals at noon on the first working day - If sales and F&I can't determine gross profit, financing, payoffs, etc. by noon

and **provide a recap and check off sheet** - then it's a "next month" deal. If you don't agree to this – then nothing works.

- ☐ You will be available to review and approve the draft financial statement from 2pm – 4pm on the 2nd working day. Please don't add more car deals to be posted at this time but of course adjustments to expenses, other income, etc. can be made. A list of posted car deals will be provided at the beginning of the 2nd working day.

Count me in! Accepted by:______________________________

Action Plan for More Time

- ❐ Figure out what you are doing all day; T, F, and M
- ❐ Determine what you'd like to do - goals
- ❐ Find those Monkeys
- ❐ Eliminate Time Wasters
- ❐ Organize your office; file cabinets, chairs in your office, inbox
- ❐ Get the approval for 4 hours free time off
- ❐ Convert to a 3-day Close

Notes:

Understanding Tasks

As a little background, I was generally a happy smiling controller with great multi-task and time management skills (look at the graphic on the cover – my granddaughter drew that of me!) I could be looking up stuff on my computer, handing someone a book they needed, loaning out my stapler (why was my office the only office that had a stapler?) and still safely take a sip of my tea…ahhh. All was good UNTIL, my dealer would either call from his 20 group meeting, or buzz me and ask…

"Are we doing this…" or "Is … being handle?

Then the fire drill would start. In a panic, I'd be bouncing around the office, trying to get the answer. It was then, that I decided that I'd keep a full list of everything that we did in our office and who was doing it – including when it was last completed. I'd also add instructions on how to perform the task and even store the diskette (it was a long time ago) that had the spreadsheet that was supposed to be used. I'd keep a copy of the most recent reconciliation or notes about the task. I kept all of this in a big binder that I called the "Red Book."

Over time, I got a little more automated and started keeping them in a folder on my computer and a master list in Excel. For now, let's analyze the concept of Tasks;

There are a few things to consider when thinking about your tasks;

- Should the task be done daily, weekly, monthly, once a year – or just during the close?
- Is there a faster and easier way to do the task?
- Is this a task that should be done by the accounting office?
- Who in the accounting office has the best skill to perform the task considering internal control guidelines?

Why Daily, Weekly, and Monthly Tasks?

You could in reality do your monthly sales tax report daily – but why would you want to? I've done it before – right after a client went on a new DMS system to make sure the setups were correct, but I don't recommend it for most Task-based offices. You can have your office reconcile your bank daily and I know many dealerships that do, but for me – our company transactions are too few to make sense in doing the reconciliation daily. When cash is tight, I might start reconciling it weekly until things get better. The general guide to a daily, weekly, or monthly task is;

- Daily tasks answer the question, are we making money? They normally contribute to the DOC – daily operating control report

- Weekly tasks can help you make more money and normally analyze items like technician productivity, closing ratios, and labor gross profit.
- Monthly tasks help us keep the money that we made and are concerned with collections, metrics, and financial analysis.

Dealerships are different; I have yet to encounter two that are the same. Since performing the task daily has the most urgency, I suggest that you start there. Look at the task and ask, could this task be done daily? Here are some guidelines in determining this;

1. Is it truly easier to perform the task daily, rather than letting the work pile up?
2. Do other people in the dealership benefit from if the task is done daily
3. Will the dealership save money if we perform the task daily?

How to Assign Tasks by Skill Levels

I normally like to break down everyone in the office into 3 skill levels; A, B, and C and then try to organize your tasks by skill level.

A or "anyone" skill level could be done by almost anyone in the office. This was normally putting things in order or filing.
B or "better" skill level was for those better qualified
C skill level meant that the task could only be done by my office "cats." A cat-like person in my office was confidential, accurate, and timely.

For example, reconciling reserves or holdback might need a better qualified or B-skill person, but I'd need a cat-like office person to accrue manager's bonuses. This task required that the person be confidential and accurate – along with getting it done on time. There is also a "D" task, that is something the dealer has required that a certain person does. For example, your dealer might only want a certain person to type his rent check, take his dog to the vet, etc…

How to Become a Task-based Accounting Office

Over the years I have found the "clogs" for our accounting office are the car deals. They hold up closing the books, payroll (sales commissions,) and cause interruptions. For some reason the flow isn't even – although the customers purchase vehicles evenly throughout the month. How can we get car deals into our office faster and get them posted quicker? It is the same way as getting service vehicle worked on – you can pay a "flag" for doing deals. Get the dealer to provide you with a fund for doing car deals and pay a bonus for how many each person gets done. I realize this is revolutionary – but wouldn't you like the whole office being able to do car deals? It would give them such an understanding on how the office works and provide a more even distribution of the work load. Shocking? Again, if you want to know if this could work – just go watch the technicians in the shop. They work as fast as they can and are always looking for another job. Wouldn't you like that the same in the office? Let's use as an example an accounting office with 200 retail deals a month. The flag fund is $5.00 a deal or $1,000. Obviously your billers are going to get the bulk of this income, but what about an accounts payable clerk that was able to do 20 deals this month and get a $100 bonus? Could you have asked her to do car deals when she got all done with entering her invoices or reconciled her statements – sure…but she's never "free" to do more work unless there is something in it for her? To get this system going, you'll need to work in phases;

Phase 1 – Get the Bonus Fund

This might be your toughest hurdle, but if your office is looking for raises anyway, this might be a way to "sell" the new plan. Ask the dealer if he would pay $1,000 more a month to have every car deal on the DOC at the end of the day, all the items due on the heat sheet, lien payoffs not missed, and payroll ready in plenty of time to sign instead of rushes to get things done at the last minute? Sure! Just ask if you can try it for a few months and if it doesn't work, you can take the $1,000 and work on old-fashion raises instead. The controller can even do car deals to earn an extra $200-$300 a month!

Phase 2 – Assigning Tasks

If everyone is trying to grab that bonus flag money every day, then you'll need to have a way to monitor their other daily tasks to make sure they are completed. You certainly don't want the person who is supposed to make the bank deposit to skip that and work on deals instead. This might be hard to do manually, but I guess if they violate the rules, then you'd restrict them from getting flag money for 30 days.

Phase 3 – Office Introduction and Training

You'll need to introduce this to the office and set a time for when it will start – let's say in 2-3 months. Next you'll need to determine what constitutes doing a car deal and create a worksheet. Then you'll need to hold training sessions where you teach everyone how to "do a car deal." The person who completes the deal then copies the worksheet/recap and turns it into the controller. You might need some rules;

1. All deals will be in stack in the office and you must take the deal from the top (no sifting through looking for ones without trades, contracts, etc.)
2. If you make a serious error on the deal; forget a step or mis-post the deal, then you lose the deal and the flag goes to the person who corrected.

3. You can go into the F&I looking for deals if the stack is empty (great at getting the deals into the office faster!)
4. You must have your daily tasks completed and any weekly or monthly that have a due date of today or sooner. If you violate this rule, you are off the flag plan until you have 30 days of 100% task completion on time.

Phase 4 – What Else Can you Flag?

If your office likes this new way to get all the work done faster and better – what else can you flag? What about warranty claims? I don't like outsourcing warranty claims – it is ineffective. Why not train your office to do warranty claims and put all the ones that need submission in a pile next to the car deals. You can pay a little less if you want – but make sure they are on paid claims. What about collections? Most collection agencies work on commissions; maybe you have some office staffers that might like to go on that plan. Be careful who you select and make sure they are properly trained. What about annoying things like cleaning out the file room or organizing the cabinets? You can use a flag system for that too! Follow-up calls and letters? Almost any task can be flagged – it is the way the shop functions, why not you?

Tasks – Daily

Most of these tasks are considered daily not because it is easier to perform them daily, but because the dealership needs them entered. It would be much easier to sit down once a month and import all the new vehicle invoices and then update to the DMS, but your sales and service department depend on this information been accurate daily. The following are only a guideline on how you should perform the task to make it go faster, but also in this section because they often qualify by the 3 items in the previous section.

Add New Vehicles - post accounting for purchase

Hopefully this process is tied to the process of adding the new vehicle into stock. If the vehicle was added when it was ordered or shipped, then the inventory clerk or biller can merely bring up the vehicle, change the status to "in stock" and update the price fields. Then if necessary, create a journal voucher to post the cost, floor plan, etc.

Add New Vehicles into DMS

The key to adding vehicles into your DMS, is to NOT add them; rather to import from your factory and then add some specific information. The more you can import – the better. As a task-based controller you need to find out if your DMS and factory have a method to import vehicles. If not, keep asking when that will be available.

Sometimes if you order your vehicle via the DMS, then you can bring up the order and then change it from order to "In Stock." The next best thing is to use the DMS VIN decoder and let the decoder fill in most of the information. If you get a shipper notice of incoming vehicles, then get them entered when they are on the way instead of having to stay late on Friday night when a transport truck arrives. Almost any office person can perform this task if you have set up firm procedures and carefully monitor them. If you want each new vehicle to have a pack of $200, then run a report monthly to make sure the entry was made – or better yet, set it up as a default. Vehicle shipping notices can be one of the most valuable things for a good Internet or BDC manager. Each truckload of vehicles coming from the factory has a shipping invoice. Look at the "vehicle desired" field in your database and see if any match these new arrivals. You can also post on your website a "newly arrived" section. The shipper is one of the easiest ways to keep your website fresh and let your customers know that you have all the latest information.

Add Used vehicles into DMS

Used vehicles that come via trade-ins should come in automatically. Bring up a trade in your F&I system and look at the information. After it gets added to inventory when the deal closes, bring it up again and make sure all the information that you need came across and if not, open up a support ticket with your DMS provider. For example if the F&I manager enters the license plate and expiration date, then make sure that valuable data ends up in the inventory file. If you purchase vehicles from an auction, then get the auction to fax or email you the purchase sheets. This enables you to start assigning stock# and entering the vehicle. The best person to enter the vehicle is the same person who stocks these vehicles in. This might not work in your dealership, but a true inventory manager should not only stock in the vehicle, but make sure the information is correct in the DMS. If you use the last 8 of the VIN for the stock#, then you don't require pre-printed stock# tags. If you worry about duplicates, then prefix the last 6; AU123456 would be an auction purchase, TN123456 could be Trade-in from the new department. If you use buyers, then SJ123456 would be a vehicle purchased by Sandi Jerome – your used vehicle buyer for Florida. A cashier can also be a good person to stock in vehicles. They often have long stretches of time when they can perform this work and since many used vehicles need to go right to the service department; they can also open an internal for the vehicle and give it all to the dispatcher. Auction purchase invoices can be one of the most valuable things for a good Internet or BDC manager. Each truckload of vehicles coming from the auction has an invoice. Look at the "vehicle desired" field in your database and see if any match these new arrivals. You can also post on your website a "newly arrived" section. The invoice is one of the easiest ways to keep your website fresh and let your customers know that you have all the latest information.

AP accounts -Add new

Many dealerships might consider this a more monthly task, but if you are using the purchase order system on your DMS, then you need to perform this daily to enable the personnel to be able to add a purchase order for each new vendor. Make sure you go through the approved vendor process to get the vendor approved.

AP invoices for New/used vehicles

If you use a tightly controlled purchase order system, then there should be no chore to entering the AP invoices for new and used vehicles. A correctly filled in purchase order converts to an AP invoices and the charge goes to the vehicle. Spend some time studying this process and find a way for the same person who issues the PO to also charge it to the correct vehicle. When the AP clerk receives the invoice for payment, they can double check that the vehicle was still in stock – or that the due bill was set up and then make any necessary adjustments per your policy

AP invoices for Parts

Again, this is a function that can be automatic. If you properly use the purchase order or stock order system, then your parts manager will have already entered in most of the information. This manager is tech-savvy, so it is easier for the parts manager to enter the cost amount and discount than to stamp the invoice and write in those amounts. Find a way for this process to happen before the invoice arrives in the office.

AP invoices for Sublet

Sublet write-offs happen because of a flaw in the process. Like parts and vehicle invoices, sublet invoices must go through the purchase order system and that system must tie-in to the repair order system. When the invoice arrives in the office, the AP clerk only verifies that the sublet was included in the repair order. Instead of spending the time entering invoices, the clerk can spend the time double checking that the process is working and when it fails; finding a way to improve.

AR - NSF checks - Adjust bank

You need to set up an alert from your bank whenever a check is being returned. Most banks are capable of this or if you have business banker, this person might be willing to email you. As quickly as possible, reverse the bank entry and get the item on the heat sheet and past due list. This entry can be completed by anyone except the person who reconciles the bank statement, the best person is the AR clerk. A good method is to have a NSF account receivable schedule that produces statements - or a series of AR account numbers. If you sort by name, start the name with NSF-John Smith. As soon as a NSF check is received, check to see if maybe the funds are good you can redeposit. If so, make the entry to credit cash in bank and debit the AR schedule. Make another receipt to "redeposit NSF" check and include the check in the deposit.

AR-Verify references on new account forms

If you can run a credit report on the owners, then that is the best place to start the approval process. If it company, then inquire about payment history, the credit limit they have set and how many returns are made. The parts department has worked hard to get this customer and now we must quickly get the references done so the parts department can quickly fulfill that first order. It might not be best to have your AR clerk perform this task, rather give it to a more friendly person like a Customer Service representative with a carefully created list of questions to ask the references. You might also make it easy and email the references your questions; do they pay on time, what is the high balance?

AR - Send completed application to management for approval

As a controller, I used to consider "letting them charge" as a gift that I bestowed up on the customer and parts manager. In reality, it is the opposite. When a parts department turns in a new account application be aware that this has taken a lot of effort to acquire the new customer. Consider having the approval process be a joint decision between the parts manager and GM, since they will suffer if the account goes bad and you have to write off the balance.

AR accounts - Approve New

Before management approves the new AR account, have a policy of your requirements; for example, never over 60 days, a credit limit of more than you're considering, and returns less than 10%. Never turn down credit without a meeting with the parts manager and general manager/dealer.

AR accounts - Add new

If you must assign the customer number (I prefer an auto numbering system) then try to make it mean something; for example BRUNT33 Might be a customer number for Brunting Industries with a zip code of 33307. Most of the newer DMS systems have a great way to filter accounts by type, so a simple auto-numbering will work.

Bank Reconciliation

I find in my seminars that about half of the attendees have switch to daily bank reconciliation. If you're trying to make the switch, there are three basic methods to use; 1. Schedule cash in bank. Log into your bank and print the transactions and have a clerk post those to the cash in bank schedule. The office debit is to your "per bank" control number and the credit and debits clear your schedule detail. 2. Use Excel to maintain your cash in bank. 3. Use the bank reconciliation software provided by your DMS system and manually clear items daily instead of monthly.

Make sure you always print reconciliation for the last month of the year – a list of all outstanding checks, deposits, and the detail from the bank for the exact same day they produce a printed statement. Your auditors will normally want to review this old fashion method.

Cash - form 8300, complete

Pull out of deals, Daily. Hopefully this form is becoming more of an F&I form that is considered on each deal and completed during the sales process rather than struggling to determine how much a customer paid total in cash and cashier checks. The F&I manager already knows how the customer will be paying, so it is easy for it to be completed then. At the time of this manual (2012) a business must file Form 8300 to report cash paid to it if the cash payment is: Over $10,000, Received as: One lump sum of over $10,000, Two or more related payments that total in excess of $10,000, or Payments received as part of a single transaction (or two or more related transactions) that cause the total cash received within a 12-month period to total more than $10,000. Received in the course of trade or business, Received from the same payer (or agent), and Received in a single transaction or in two or more related transactions.

Cash - form 8300 Submit

At the date of this manual (2012) a business must file Form 8300 within 15 days after the date the cash was received. If there are subsequent payments that are made with respect to a single transaction (or two or more related transactions), the business should file the form 8300 when the total amount paid exceeds $10,000. Each time the payments aggregate in excess of $10,000 the business must file another Form 8300 within 15 days of the payment that causes the additional payments to total more than $10,000. Make this an item on your deal checklist.

Cash Flow Analysis

A Daily Cash report is a good way to determine your daily cash flow. I like to also think of this as a float calculation. Each bank and dealership has a different amount of processing time – or float. This has decreased during the years as more payments are via EFT, but you should still know that amount. That way, when your cash per the computer goes negative on the DOC, you can assure your dealer or GM that you're okay with that. Doing your Daily Cash Report helps you understand how much deposits should be; how much payroll and payables drain your cash and what reserves you have like CIT and VR(Contracts in Transit and Vehicle Receivables.) It is also a warning if CIT and VR is high one day – did the clerk pull the contracts this week? If your deposit is low, did they deposit all the checks - or apply EFTs? If you have an ADP system, use the M.I.S. special "Cash Flow Format Columns." Create and set up detail lines for all balance sheet accounts, taking care to group the accounts into lines of like categories. Then set up your report as a custom format using the special "Cash Flow Format Columns" of your choice of

170A through 174C. For a format friendlier to non-financial managers I like to set up the report to print the line numbers, download the report into Excel and use the line numbers as the key reference item in Excel's VLOOKUP function, allowing for positioning the key data any place on the report you choose. I like to use ERAAccess query to generate the required details from the G/L and once downloaded into Excel use VLOOKUP to automatically generate the OCF report or a sources and uses of funds report. From ERALink's Run menu choose Query Builder. Select LIST as the Verb and for File choose GL. Choose the starting and ending dates for the Items and for Output select "ACCT.NO ACCT.DESC BEG.BAL CUR.BAL END.BAL ACCT.TYPE." With your ERALink Query set up to automatically open and load the Excel Worksheet with the results of the report just run the report and copy and paste the results into your pre-designed report. With R+R Power (UCS) you can create customized printed or on-screen reports to take your dealership's financial pulse. To create a customized cash flow report, open POWER Workstation, select Go To,Application List, Options,Management Accounting. Then, from the Functions menu, select DOC Control Options, Define DOC Report Option. Select a report to modify or close the DOC Report Lookup window and press Ctrl+A to add a new report. Enter a report number and press F12 to add it. Then, enter the report Name, enter at least one keyword for the information you wish to include (i.e. CMNET for current month's net change), and press F12 to update the report. In the Define DOC Content sub screen, add all keywords for the current and previous account balances you wish to pull to the report and press F12 again to save your changes. Once you have added all keywords, specify the details for each line of the report in the Define DOC Line sub screen, including G/L account numbers or ranges. When you have completed your report, you can request it in Print, Standard Reports,Printed DOC Reports. (If you need help setting up this report, contact your client support representative). With POWER Workstation drill-down capabilities, you can view report details on screen by selecting Instant Doc from the Accounting Functions menu in the main MAC application. You can use the F9 key to drill down to an itemized listing of accounts composing a line of the report. Pressing F9 again will display the G/L details for an account, where you can view individual posting details, if desired.

Cash Sales – Reconcile

This should be done daily, but are your cashiers coming up short each day? Does it take hours for your staff to reconcile cash sales? Does that account never zero out at the end of the day (or month?) Here are the most common reasons why and how you can stop it. 1. They don't know how to give change – or get too busy to calculate it properly. Why not get an inexpensive cash register from Costco or another discount office supply? They can ring up the repair order number, indicate the payment type, and if they receive cash, it will tell them how much change to provide. I know that you think that anyone can give change, but studies show that the error rate can be as high as 25%! 2. Your credit card payments are not cash. If your bank does not give you 100% credit for your credit cards on today's deposit (with a month end discount fee) then your credit card payments are not cash. Even if they give you the credit in a few days, that is not cash. Set up a clearing account for credit card payments – either a cash account or an AR account and change the payment type to a charge account. 3. You cash checks or give change to employees. The cashier's are not a bank and they should not be cashing checks. Think about the 25% errors in item number one and apply that to a large check cashed for an employee. In addition, it requires you to keep a larger cash drawer. If change is an issue, buy a change machine or switch to vending machines that take larger bills. If you allow check cashing, you'll never be able to perform a cash in – cash inventory audit (see item #6.) 4. You use the cashier drawer for petty cash. Don't allow this – once a cashier sees how much is taken out to reimburse lunches – it doesn't take them long to wonder, "why not me?" Only thirty percent of the general population is honest – so if you provide the opportunity – and the justification (everyone else is doing it) then you're encouraging and training your cashiers to reimburse themselves. 5. Make sure the cashiers indicate on EVERY repair order and ticket the payment method – check, credit card, charge or cash. 6. If you think your cashiers are stealing, the best way to prevent this is to install a video camera pointed at the cash drawer. The monitor is put in the dealer's office. Of course he can't watch it all day, but the simple knowledge that the transactions are being recorded discourages theft. It's a serious solution and rather, I recommend buying a simple electronic cash register as the first choice. What about enabling cashiers to adjust tickets? If you let a cashier adjust a repair order, you are providing another opportunity to steal – but I'm actually okay with this. You can run a report and find any repair order where the shop supplies were removed or items adjusted– so why ruin the cashier's ability to provide faster and better customer service when a repair order requires adjustment? Make sure you have a policy to use this feature and regularly run the report and discuss items on the report with your service manager and cashier. It is devastating to find out that a cashier is stealing, but even more frustrating when you can't figure out if they are stealing or just incompetent. Try some of these methods to isolate your issues and close the door on cashier theft.

Cashier, relieve for breaks

Nobody in the office likes to relieve for breaks, but if you rotate the task each week between 2 members of your office staff, then acceptance is a little easier. The best person to do the relief is the person who does the bank deposit daily. They would then transfer that chore to the other person on your staff that did the relief last week. This type of cross-training and check not only makes sure that policies are being followed, but also creates some empathy. For example, I once heard the person who was doing the daily deposit required that the cashiers make a photocopy of each check and then stamp the RO and enter all the information from the check onto the front of the RO. That ended when that person had to relieve the cashier and found out how hard that is to do with a line forming of customers who need a cashier. If you have a warranty clerk, this is an ideal person to relieve for cashier breaks and might be a good person to actually supervise the cashier since they know how repair orders should be completed.

Checks - on demand

If possible, demand checks should be created by the accounts payable clerk unless it is for an employee. This person is the most qualified to issue checks and already understands the 1099 requirements. The payroll clerk should issue all employee check since they will have to deal with taxing the amount later. Most accounting offices end up with check requests going to everyone – and everyone typing checks and various time and messing up the printer. If the parts manager needs a check for a purchase and they go to accounts payable, that person might say, "We already have an account for this company" and save a check.

Create daily receipts

Receipts are best done by person receiving the payment. The first step is to make sure you're using only the cash receipt program that comes with your DMS system. I still find receipt books hiding out at most dealerships. Sometimes they are in the bottom drawer of the F&I managers' desks, or at other times in the sales office. It is time to round them up and stop using them. Not only does it enable fraud (by giving a customer a receipt and not creating a duplicate or turning in the money) – but they cost money. All it takes is for you to train your F&I manager how to write a receipt with your Cash Receipts program. It will save you money, increase productivity (the office doesn't have to create another receipt) and provide better internal control. Round up those receipt books today and put them in your office. Make an instructional sheet on how to write a receipt, and the next time someone is looking for a receipt **book, hand them the instructions instead!** One of the biggest horror stories that I've heard about depositing checks is about a clerk that got a little behind in her work. The one thing that she just couldn't get to was receipting in the checks. It was such a cumbersome job that instead of getting to it daily, she'd paper clip the checks together and carefully lock them up in her top middle desk drawer. One night she left the drawer unlocked and the controller was looking for a pen....well you can imagine what happened! He totaled the checks and they added up to over $250,000! This was also at a time when the dealership was running out of cash. I will agree that receipting in checks is a cumbersome job. You have to find the account number and control number for every check. If you are utilizing open item AR, you also have to enter every invoice that is being paid on the check and often times it doesn't balance. What is the solution to this? Consider a deposit clearing account. Most higher level accounting systems use this method. All checks are receipted in when they arrive at the dealership, but let's start at the beginning to make a better and faster way to receipt in checks. First you need to start at your mail system. My favorite system is where the mail is given to the receptionist. This person quickly slits open all the mail (except items that are required to remain sealed like a personal letter to the dealer.) As she is cutting the mail open, she puts any checks into a pile. When she has all the checks, then she ignores the rest of mail while she receipts in the checks using the cash receipt system. There is also nothing worse than opening up a car deal and finding a check. The reason this happens is that we've made it too hard to get them receipted in. If you train your receptionist to receipt in checks, then they can handle these checks the same as their "mail checks" and be ready and available to issue a receipt. At some dealerships the only way to get a receipt is to either wait in the cashier line (horrible at 5pm) or get someone in the office to do it for you. Even worse – the F&I manager issues a "receipt" and then dumps it into the nighttime safe. Don't get me started on the stories about these receipts and missing money. If instead, the receptionist can receipt in checks – then there will be no need for a sales receipt book – since the receptionist should always be there when the dealership is open. In order to make this system work, there must be either a way to "apply" amounts from the deposit clearing account or edit the receipts. This method will vary by each DMS system, but if you can get it to work, then you'll avoid having the dealer find stacks of un-deposited checks and improve your cash and internal control. The biggest issue you'll have is who will be responsible for clearing the deposit clearing account? You might want to prefix the posting description with AR, CD, and MS – for Accounts receivable, Car Deal, and Misc. Whoever would normally handle these types of checks is also responsible for applying amounts in the deposit clearing. Yes, you might be making a mess, but it is minor compared to running out of cash or not getting checks cleared in time. The key is to design a system that works for you.

Close repair orders – Cash and charges

The best time to close a repair order is immediately after the customer has paid and by the cashier. Some DMS systems enable you to "turn this on" and others have this as a normal feature. The cashier should quickly make a few "checks" of the repair order BEFORE the customer arrives; does the labor match the parts – for example if there is **a labor line that replaces a water pump and a part line with a water pump? Are shop supplies correct? Is there a technician for each labor line? Is there a cost for each part number? If so, then print the customer copy and have it ready for the customer BEFORE they arrive. When the customer arrives, stamp their copy paid (if that is your policy) and enter the payment information in the computer. Print the accounting copy (or better - save for archive) and post the customer pay portion of the repair order.**

Close repair orders – Internals

Internal pay on Customer repair orders should also post immediately when the customer portion is closed – on a Daily basis and by the cashier. The internal pay amount must have already been pre-authorized so it should be easy to pick the pay account (policy, due bill, etc.) and close that section of the repair order. It makes no sense to get the amount authorized – the work has already been done, so charge it to the account that makes sense. For example if the internal labor line says "fix oil stains, shop policy" then it is easy to pick a payment code that is shop policy. If it says, "fix dent, charge to new cars" then charge that payment code. Sending the repair order to the new car manager for approval AFTER the repairs were done is non-productive. When they analyze their expenses later and find this was an error and should be charged to the shop, then they can fight it out between themselves – but in the meantime, the repair order gets closed and accounting is complete.Here are my tips for closing internals quickly; Create service pay codes for each vehicle inventory account. Require that advisors put in the complaint – who is to pay for the work ("per due bill", "new car policy per Jim", etc.) Close all internals to vehicle inventory pay codes if you don't know for usre.Teach your cashiers or service advisors to close each night before they leavePay a bonus for each day that there are no open and completed internalsGive a copy of the closed internal to your biller to move if necessary to due bills or policy. If a cashier doesn't know for sure, then close the internal to either new or used inventory if the repair was for the sales department. I know that it is frustrating to have debits hit inventory accounts rather than the due bills, but what is worse – having to write a journal voucher or repair order adjustment – or missing the $700 internal when you're selling a vehicle? Step 3 is so simple that it almost seems silly. Let's go through the process. I'm a cashier and receive a "hard copy" of an internal repair order – or a service advisor who gets notified by dispatching or the technician that the work was completed on an internal. That person closes the Internal to an inventory accounts. Of course even the most savvy service advisor will sometime close a new truck to new car pay codes, but most DMS systems will find that cost as long as the stock number is correct (a good argument for using the last 8 of the VIN!) Step 4 and 5 takes some understanding of your current process. For most dealerships, they send a copy of the internal out to the used vehicle manager for approval. That is a bad process. The work has already been completed, so no approval is necessary. They might instead send the internal to the biller to get the right accounting. Again, a bad process. In the day or days that it takes to get that person to "code" the internal, the vehicle might be desked. Let's say that the cashier closes the internals quickly to a "clearing account" on a Friday afternoon. The biller has gone home for the day and the $700 internal sits on her desk. This Used Vehicle is sold over the weekend and the desk manager had no idea the $700 cost happened. It doesn't show as an open internal and the biller has not moved it from the clearing account to the vehicle inventory. Of course many of you can say that the manager should have known that work was done – he requested it! This is asking too much of our talented desk managers who need to be focused on desking the deals, not memorizing the work they authorized. If instead that internal had been closed Friday afternoon to the inventory account, it would be in the cost of the vehicle and displayed in the detail. Yes, on Monday, the biller might have found out that that the $700 was for a vehicle sold last month and it goes to either policy or a due bill and now she has to journal voucher it off – but how much does that cost? When the opposite is a true loss of gross profit, then let's focus on making money instead of having clean books! I've mentioned before a bonus for the cashier or booker for having no open and completed internals at the end of the day (send me an email if you want this information) but what if you don't have cashiers – and advisors close the tickets? That is easy. I know an office manager who if she finds any internals that are completed that haven't been closed from the day before, then she change the advisor number to her own and closes it. She pays herself the service advisor commission (with the dealers blessing and encouragement!) Do this a few times a month and show them how much it costs them, and I bet advisors will get them closed.

Closing Ratio report

This is probably the one report my dealer wanted to see first thing in the morning and every day. The receptionist is the perfect person to do this report because they know how the "up" process functions and is often a part of it – including a login to the CRM portion of your software. Ideally, this report should be emailed to each recipient and I like to include each salesperson in the distribution list. People are naturally competitive, so why hide this information from salespeople?

Contracts in Transit/Vehicle Receivables - Schedule clean

If you download the schedule to Excel and filter on any credits and investigate. They might be down payments in the wrong control# if they are older. Next filter on small debits. Are they mis-postings, less than $1.00? If you have multiple schedules, download together and look for errors on the account, control. Be sure and correct the process – why don't the clerks know the correct account for down payments vs. payments on contracts? Create a cheat sheet if you need to.

Create a jacket for trades

If you are still a non-paperless why maintain the inventory jackets? Everything in the jacket is in the computer. The jackets are expensive to order and expensive in labor to fill in the VIN#, year, make, etc. Now that we keep the MSO and title in a fire proof safe in a binder, (I hope!) you don't need to maintain these jackets. You are especially causing a lot of extra expense by making copies of the repair orders, bills of lading, and invoices. Think of all the expense to then enter all the charges to the vehicle and keep a running total – the same thing you'd see if you looked up the vehicle on the computer. That's a triple savings to toss away an old manual process and bring your dealership into the paperless age! But if you can't get away from having those jackets, then it should be the future deal jacket and the F&I manager should complete it and turn it in with the deal. It only needs the stock# for now.

Create daily bank deposit

The daily bank deposit should be done by a **person who does not reconcile that bank statement nor does the collection and manages accounts receivable.** If your bank requires a manually created deposit slip – she enters every check and amount into the deposit slip. If you haven't been using the cash receipts system all day, then she brings up the cash receipt system on the computer and receipts in every check to a deposit clearing account. She enters the receipt number on the check and scans the checks into a PDF file saved in a special folder on your network. The file is saved with the name deposits121509sj – with her initials at the end. She totals out the deposit slip and also scans that too. If your bank instead has provided you with a check scanning system (and you don't pay a premium for each check processed) then instead of creating the deposit slip, she would use the system to scan, and then print the deposit report. She then uses the same steps as above to post and scan. The objective is to get checks into the bank as soon as you can. Nothing is worse than getting an NSF check that was given to the dealership on a date and you don't receipt in for a few days and then don't deposit until the next day – so 3-5 days went by and you didn't know it was going to bounce. When she comes back with a deposit slip, then she can enter the receipt lines and close the receipts and deposit.

Credit Life AH policies - Refunds, process

You should always get a written request for a refund of a credit life policy. Processing these daily when requested means that you'll get your refunds faster and that will make the customer or bank happier. I prefer to have this performed by the F&I office since they will be suffering the chargeback.

Credit Life AH policies - Claims

Keep the claim forms at the receptionist desk and have this person send out any claim forms or route a customer to this desk to obtain them. Having a printer at the receptionist desk with a copier is a great idea and then have the receptionist make a copy and mail in the envelopes kept at her desk.

Credit Life AH new policies- submit

Normally it is good to send these in daily and should be submitted by billing or contract desk clerk. The LAH company normally supplies envelopes (ask if you don't have) and just put the policy in there when the deal is finalized (proceeds received.) To avoid cancellations and rewrites, make sure the deal has been paid before you put it in the envelope. You might need to adjust this if your LAH provider requires they be sent in sooner.

CRM - provide a Daily Work Plan to salespeople

This might be an odd thing to see in a task list for the office but I find that most CRM software fails because there is no one to run the software. Trying to make a sales manager operate the CRM software – at least the follow-up portion – it is asking a lot of the type of person that profiles out to be a good sales manager. The key is to find out if this is being done and who is the person generating it. Often the software provides the Daily Work Plan electronically directly to the sales person as a feature and if so, then you need only check on it monthly to make sure it is still functional.

CRM- Make follow-up calls with non-buyers

Again, this is another item that might seem strange to be in this book, but it is no different than making sure someone is submitting warranty claims. Find out who makes the calls and what type of record is created so that you can do a spot check and make sure it is "being handled."

Daily cash sales - make sure it is zero

Probably one the easiest things that I did each day was perform an inquiry to the cash sales account and made sure it was zero. If you are posting repair orders as they close, then filter for all detail prior to today's date and today's date for the receipts. The total of this detail should be zero – no excuses. If someone posted a repair order or parts ticket to cash sales and the check is coming in tomorrow – then it is not a "cash sale" – it is a one-time charge. If you don't let this account get away from the zero balance, then you'll never need to reconcile.

Dealer trades – get paperwork ready

Doesn't it always seem like dealer trades happen every Friday at 5 pm? Actually they happen all through the week and day, but the ones we remember is those that have to be done when everyone is trying to get out the door for the weekend. To make the process go easier, make sure every person in the office can prepare a dealer trade and that there are clear instructions on how to perform. A worksheet provided to the sales office makes it easier. This worksheet should show at the top the vehicle that is being traded and at the bottom the vehicle that is being received. It should indicate who initiated the trade (their trade or ours) and who approved it. The worksheet should also have the amount of the check if required or the amount of the check that the dealership expects to receive. I do not like swapping checks – it is rather wasteful to create two checks. When the paperwork is complete, put it in an envelope at the receptionist desk so that is the first place the driver looks before disturbing the office.

Dealer trades - payoff Floorplan

When the dealer trade worksheet is complete and a check has been receipted (unless the floor plan was a swap,) then attach the receipt or check copy to the worksheet and give to the same person who pays off floorplan daily to include with that day's payments.

Dealer trades – Add to Floorplan

When the dealer trade worksheet is complete and a check has been receipted (unless the floor plan was a swap,) then there should also be a copy of the invoice so that the new unit can be floored. Even if the vehicle is sold, it should be floored to qualify for the insurance that is often included with floorplan agreements and in case the deal doesn't go through.

Dealer trades – post

The easiest way to post the dealer trade is when you write the check or receive the funds -- if you're using a difference check (again, I don't like swapping checks.) Instead of using a clearing account, debit the amount of the vehicle you're receiving and credit the one that is being traded. The difference is your check amount or receipt. I realize that it might be the policy in your area to swap checks, but just say that your company is now doing difference checks and start a trend!

Deals not in Office Report

Most DMS systems update the status in F&I as "Final" when the deal has been posted to accounting. In order to get the deals not posted in accounting, you'd run a report of all F&I deals, using the Status field. Use the AutoFilter, Data, Filter, AutoFilter – Next to the S or Status field – select B for example for ADP deals and the status code or name for other DMS systems for booked deals in F&I (not finalized in accounting) and P for Pending deals. When creating a Pending deal list, move to the Buyer-Name and select –Non-blanks. You can also use the AutoFilter and select where the status does not equal F (for finalized) - or whatever status code is your DMS system's code for finalized deals.

Heat Sheet

Are you using Excel for your Heat? Are you entering the data manually? With a few modifications, you can make this daily chore a little easier by downloading from your DMS system (unless you're using one of the newer ones that have a built in Heat Sheet.) In addition, you can provide more information by using the VLookup function. By making the heat sheet easier, you can switch to a daily heat sheet instead of a monthly. First of all you need your F&I managers and sales managers to enter their notes about the deal in a field on a deal screen – and not in an Excel spreadsheet or on a manual log. If you're going to create the heat sheet daily – then you won't have the time to type in all these notes over and over again. Find the field on your DMS system where these notes are stored, you'll need that in the next step. You'll need to download two reports; first would be your summary schedule (mini-schedule) of all the accounts that you want to include on the heat sheet. This needs to be a one-line schedule. Cut and paste this into Excel and clean it up. Next you need to run a report from your car deal file of all the closed deals for the past few months (or the oldest date on your schedule.) You'd use RPX on ADP, ERAAccess for R+R, a report of the FRM in UCS. Include the F&I manager, the bank, the sales manager, salesperson's name, the notes and any other information that is not on your summary (mini) schedule – but would help the managers collect the funds. Copy this sheet into the same Excel workbook that has your schedule. Now create a few columns to the right that use the VLookup function to populate the valuable information from the car deal – like the bank, F&I manager, and especially the notes. Reverse the process for the car deal list and find any unmatched posted deals – and these are the deals not in the office. Email this spreadsheet daily to each manager. The total time to do this heat sheet is less than 15 minutes – once you get the hang of it. You can speed up the process by "linking" to the files and right clicking and !Refresh. By creating it daily – it becomes accurate, timely – and you can give it to the people who can get you results!

DOC/MIS – distribute to Managers

I believe that with today's technology, each manager should run or preferably view their own DOC. To print out DOC for each manager is wasteful. If you must print a copy, this is a good project for the payroll clerk or receptionist since they usually know which managers are here this week.

DOC/MIS - prepare

I recently visited a dealership group that had one of the most detailed and extensive DOCs that I had ever seen. Many of us ignore the DOC or MIS report. It can be ugly, inaccurate, and just plain boring. For this dealership, the DOC is truly what is the acronym stands for; a Daily Operating Control. If you are going to depend on this document to control your dealership, then it needs to contain certain features to make it useful. For those of you on ADP, this is called the
MIS report – or Management Information System. For the sake of this tip, I'm discussing the MIS and the DOC. Let's go through some steps to make your DOC the money-making tool that it was designed to be. Step 1 – It must be right. Somewhere on your DOC there needs to be a summary section that has, Total Dealership Gross Profit, Total Expenses, Total Other Income, and Net Profit. These four lines must match your financial statement. If they don't, then you have some utility work to get started. Over the years you might have added account numbers to your DMS system without taking the time to add them to the DOC. It's time to get this document to match the financial statement. Some of the DMS systems have tools to help you find account numbers on the chart of accounts that are not pointed to the DOC. Many also have classes on how to create or modify the DOC. It is worth your time to get that DOC accurate so your managers can depend on a daily report instead of waiting for the financial statement to discover problems in their department. Step 2 – It must have pay plan numbers on it. If you pay a manager on adjusted selling gross then that line needs to be on the DOC. If they get a bonus for average repair order amount or customer satisfaction scores, why not put those on

the DOC? Average repair orders is easy, but what about CSI scores or hours in the shop? The DOC can include any GL account, but those are not accounting numbers. You're going to have to create a memo account for these items. Before you get out your DMS manual, look to see if one already exists. For many of you, these are already statistical items for the financial statement and are probably there. You're going to now post to them more regularly and put them on the DOC. The best part of this is, if you watch these stats daily, then there is no need to fix them at month end as part of your financial statement process. Anything you can do to move the month end chores to daily or weekly is great! Step 3 – It must have your dealership's "metrics" on it. What are metrics? Some of you might call them benchmarks, but I like the term metrics. Let's say that your customer labor gross profit rate is 75%. That percentage needs to be on the DOC. It's easy – just divide your gross profit line by the sales line. Other metrics are gross profit averages for new and used, F&I income per vehicle, advertising per vehicle sold, commission expense. Step 4 – Include controllable expenses. How can you ask a manager to watch advertising and commission expenses if those lines are not on the DOC? In addition, they must be posted daily. Yes, I know a lot of expenses don't happen until the end of the month, but many can be entered daily. Make sure you accrue your commissions on each car deal as it is posted. Require managers to issue purchase orders on the DMS system for advertising when they create their budget – or order the advertising. Don't wait for a bill from the newspaper. Require other advertisers to send daily or weekly invoices so you can get them posted. I know of one controller that posts the floorplan charges weekly – but that might be asking a little too much (although the amount is pretty easy to calculate or do an inquire online.) What about spiffs? Every time you issue a spiff check, debit the expense account. I know you like to debit the employees' AR account, but debit the expense account, the employee AR account and credit a controlled accrued payroll account. Change your payroll distribution to hit this accrued payroll account for spiffs and you won't have a big chunk of spiff expense at the end of the month. Your GM or dealer will see those spiffs go out as fast as you do! Step 5 – It needs to be a "state of the store" report

There needs to be other health items on the DOC. I don't' mean the dealer's blood pressure reading – but items like Total New Inventory, Total Used Inventory, Cash in Bank, Contracts in Transit, Work in Process, Warranty Claims. I know that I recommend that you have some of these items on your Super DOC, but that report is for the dealer, GM, and controller. I suggesting that you put these items on the departmental managers' page. The service manager needs to know how much is in outstanding warranty claims and work in process. The parts manager needs to know what is his inventory balance per the GL and the AR balance. The new and used vehicle managers need their Inventory Balances, Contracts in Transit and Vehicle Receivables. How does the DOC increase profit? Managers want to do a better job of running their departments – but unless we point out the places they can make improvements, they end up being reactive managers. Reactive manager "act" each day based on what hits them (angry customers, showroom ups, phone calls.) Proactive managers see a need (higher gross profit percentage) and then make a plan to increase that metric to your guide. This could involve hiring some lower priced technicians to do LOF work or developing a better menu mix of their service offerings. If you have a profit area that you want your managers to work on, make sure it is on the DOC!

DOC/MIS – Review

If you watch the "Day" column of your DOC and see a day where you didn't hold your metric, and then ask to see the detail for that day. For example, I know a controller that saw a drop in parts retail from 32% to 12% in the Day column. When she detailed this data, she found a large parts ticket sold at cost to a technician. Digging deeper, this part went onto another employee's car and the labor was cash under the table. Ouch! This variation would have been lost in the end of month financial statement but sticks out in a daily report. Metrics can be entered into your forecast or budget columns. Just do the same calculation in the forecast (dividing your forecasted gross by your forecasted sales) and you'll have the benchmark guide for each metric for your dealership.

Employee Add to Payroll System

After receiving a completed payroll folder from the receptionist, then add the employee to the Payroll system, obtaining a new employee number. Try to make sure the employee, salesperson/advisor, and customer number all match.

Employees – Make new Hire Packets

The receptionist should have a supply at that desk of blank employment applications and new hire packets. This keeps the traffic of coming into the office a minimum, especially around the sensitive information on the payroll clerk's desk. Managers will get an application, packet from the receptionist and also turn completed ones back in to the receptionist – who will then take to the office during the next break.

Employee insurance - add new employees

After preparing the month end report, determine for each employee their insurance start date and put their insurance application in the folder for the company that they picked with the date at the top of the form. When it is time to pay the insurance company, see if any employees are close to their start date and mail that application with that month's payment. Add the deduction to the payroll system.

Employee insurance - delete terminated/Cobra

Each week find any employees that have been terminated and prepare the Cobra forms and notify the insurance company.

Employees - Interviews

Personnel is one of your biggest expenses and hiring bad or the wrong employees cost you profits. Our industry has never been good at reducing turnover. Hiring and training new employees is expensive. In addition to these personnel costs you also have the direct cost of running an advertisement, printing job application forms, performing background and drug tests. In most dealerships prospective employees must be interviewed by two managers. That interview time is wasted time – it doesn't make any money. I will admit that managing employees was my weak point. I was great with numbers, the computer system, and understanding how to increase profit – but when it came to supervising employees – I hated it! Sometimes it was just easier to do everything myself (sound familiar?) Fortunately, I went to some of the Super Controllers that I know and researched this task based on their wisdom and combined this with my own experiences. Let's start at the top and how do you find the right controller? The number one reason why I was a successful controller was my office manager. Having a second in command is important if you're going to grow as a controller. I find the best controllers or office managers come from within (that's where I came from.) That means that for most of your staff, you're looking for future office manager potential in the interview. I've hired staff that wanted to be nurses, teachers, and actors, but they were normally for positions like receptionists that I was in a hurry to fill. Always ask their career plans.

Employees - New employee orientation

It is important to have a notification on the new hire form of the employee's first day on the job and arrange to provide new employee orientation, including completion of all required paperwork and dealership tour. This doesn't necessary need to be done by the payroll or HR manager – especially if that person doesn't have the required personality or time to perform a great orientation. You can train almost anyone how to perform this and why not try to find someone that is "up" on the dealership – like a customer service representative or receptionist? My favorite person is the dealer or GM because I worked in a few dealerships where I never met this person, or only met the person after I quit!

Employees- Recruit new employees, place advertisements

The best place to recruit new employees is your own dealership. Let's use finding a new office manager as an example. Here are some steps for internal recruiting; 1. Quickly identify anyone who has potential on your current staff. 2. Hire only those with potential for any new openings. 3. Require that each current office manager or controller have a 2nd in command. 4. Make sure you know the career path for each of your employees. 5. Make sure your employees know your plans for their advancement. This process can be used for any position. We've learned from the hiring of technicians that the best come from the apprentice program. Think of your office as a training program for controllers and office managers. Find those who have potential and give them more tasks to do each month. They want to learn more. How do you identify potential and hire only those with potential? You're looking for people who are; 1. Smart. 2. Make few mistakes. 3. Quick (to learn and do their job). 4. Helpful to others. 5. Always asking for more to do or learn. 6. Loyal and trustful. Smart is like location in real estate – the three most important things are; smart, smart, smart. Yes, the world needs file clerks too, but not as your super controller or office manager. These two positions must be smart. How to measure smart? Ask to look at high school and college transcripts. You might want to get legal advice about requesting transcripts – since it indicates age, so instead put the GPA on your application with a note that it will be verified before employment (you can require the transcript after the job offer to verify.) I included my own transcripts with any job that I applied for and I think it helped me get hired. What if you don't have anyone with potential and need to hire someone with potential? Here's a perfect recruiting advertisement – you can run in two areas – Accounting Manager Trainee and Manager Trainee Manager Trainee (Also run as "Accounting Manager Trainee"); "We are looking for a management trainee for our accounting department (or dealership.) No experience is required, but you must like supervising an accounting staff and managing a computer system. Accounting (or Automotive/Business) experience or (a business-focused) education is helpful but not required if you like working with numbers (people,) learning new things, and working with other managers/computers." What do you do with this person? Are you kidding? Every accounting office/dealership that I visit is buried. There is always work to be done. This should take a year and you need to spend serious time with this person teaching them the business. Ask a cashier during a review "what other position in the dealership would you enjoy doing?" If your cashier responds with service advisor or greeter, then you don't want to spend a lot of time training her on how to clean the sublet schedule. Instead, let the service manager know so he or she can expand the cashier's duties like setting future appointments, making follow up calls and performing service campaigns. If instead your cashier wants to be a payroll clerk or wants your job, then start giving her schedules to clean, accounts to reconcile, or show her how to break out the service department expenses for the manager each month. After you have that conversation, make sure your cashier knows that you will do your best to move her to a new position as openings come up. Often an employee will stay with you longer – even if they don't like the job, if they know that in six months, they will get moved into the office or out to be a greeter. Spend time with your employees and find out their potential and everyone will be happier.

Employees - Review all company memos to employees

I used to complain when I worked at a dealership group that we were getting "memo'd to death!" It is a good idea to have a "gatekeeper" to send all your memos through and then let that person decide who will get the memo and if it is written okay. Sometimes a memo gets written in the heat of the moment and comes across too harsh.

Employees - Terminations and Exit Interviews

An Exit interview is almost as important as an original interview or orientation. It is important to find out why the person is leaving to help avoid turnover in the future. I have found about theft, unfairness, dishonesty and other shocking events during exit interviews. The GM is the best person for this unless the person worked directly for the GM. It is also important to perform a termination checklist; remove logins, passwords, get equipment back like demos, phones, keys, and go over some of their pending tasks.

Employees – Screening tests, perform reference checks

This is a rather tedious task, but important. When you have a full HR department, it is rather easy for them to perform the task, but to squeeze this in during your normal payroll routine is tough. It helps to have an outside service that performs this function to insure quality control and this is a growing industry.

Extended Warranty New Policies - Submit

Normally it is best to submit these daily and should be submitted by billing or contract desk clerk. To avoid cancellations and rewrites, make sure the deal has been paid before you submit especially if the leinholder might be on the policy incorrectly.

Extended warranty policies - Refunds, Process

You should always get a written request for a refund of a warranty. Processing these daily when requested means that you'll get your refunds faster and that will make the customer or bank happier. I prefer to have this performed by the F&I office since they will be suffering the chargeback.

Extended Warranty Claims - Submit

Normally it is best to submit these daily and should be submitted by warranty clerk. Many people mistaken these as an accounts receivable function, but due to the complexity of the claim, especially when the extended warranty company is the factory – the warranty clerk should submit these. In addition, they should have their own schedule – separate from the AR schedule and one for each company.

F&I night deposit

Daily by same person that does the bank deposit. Hopefully you agreed with my task tip on cash receipts and have trained your F&I managers how to issue receipts using the cash receipts system instead of a manual receipt book. Not only do these have a huge internal control issue, but they are costly. Your F&I managers are some of the most highly trained users on the DMS system, so set them up to write receipts on your DMS. Each morning, run an inquiry for that "cashier" and reconcile to the funds in the night drop. It is easier than you think. You'll save time and money. Make sure your F&I managers have a night safe to put their deposit. It is important that manager never be allowed to take the money home. A night safe is normally a one-way drop into secure safe.

Factory Communicates retrieve

This is best done by the new car biller in the morning or the person who submits for rebates and also the new car managers and fleet manager. That person can read the notice while distributing. Many manufacturers now send an email notifying users of new communications that are of interest based on the user profile. Make sure one person in your office retrieves all rebate information for the master rebate book.

Factory Mailbox – distribute

Many manufacturers now send an email notifying users of new communications that are of interest based on the user profile. If that is the case, then consider stopping the old method of printing and distributing the mailbox communications. Users can now get the alert, then log into the factory website and read the mail that concerns them. If you do have to still print and distribute, the best person for this is the same person that sets up rebates, etc. That person can read the notice while distributing.

Filing – Archiving Schedules, Journals, and GLs.

Some the DMS provider do not enable you to get the archived document back out unless you bring each one up individually and print to a PDF file. Avoid this! To test an archiving system, ask that you have all PDF file from the prior year provided to you on a CD/DVD. If your provider cannot do this, then develop a different archiving system that you control. See my tips on scanning. Each month, print your schedules, journals, and GLs to a PDF printer (install BullDog it's free!)

Filing - Parts Tickets

If your DMS stores a PDF copy of the parts ticket, then you might only want to have a copy filed by the parts department and the answer to who should file is easy – the parts department! In fact, they really don't have to file, instead have them label a recycled box each month with the start date and let them just drop them in. When the box is full, enter the ending date and store upstairs. If they need a copy later, then they can find the box and yes, they might be slightly out of order, but it sure beats filing. When your storage time is up, then destroy the box. You no longer need an accounting copy if you can view the ticket online and run a report to reconcile cash sales instead of using accounting copies. It really is easier!

Filing - AP folders

It helps if the AP clerk can have a 4 drawer file cabinet next to the desk. In this cabinet should be a folder for each AP vendor for the year – and the copies of the checks in each. The old method of making a new folder for every AP vendor for every month is wasteful. Only a few vendors have so many invoices that they can't fit all the year's check copies and invoices. A better method would be to scan an invoice and then destroy. A copy can always be obtained from the vendor. I realize that we're not quite there yet – but consider this area first in your paperless plan.

Filing – Checks

The accounting copy of a check should be returned to the person who requested it and then filed. The person who requested the check normally knows the best place to file them – and better yet, have the file with the check. For example if you pulled a car deal to type a refund check, attach the check to the file (keeps you from having to pull it for the signer if they have questions.) Then when the check is signed, put the accounting copy in the file and put it in the car deal drawer for the month. There really should only be 2 copies of checks; the original that goes to the payee and the copy that gets attached to the documentation. Since we can now see all check info online with our DMS or at the bank, the 3rd numerical copy is usually not required.

Filing, Misc letters, etc.

It is good to have an A-Z file in a file cabinet *outside* the controller's office that contains files not related to customers (letters regarding customers should be in their car deal file.) This would hold files regarding taxes, attorneys, associations, and other matters. Most anyone can file this information, but anytime that you're having difficulty with another project, get up and do some filing and often the solution will just pop in your head!

Filing – Payroll

Due to the sensitive material in payroll information, this has to be done by a payroll clerk. This clerk should have a 4 drawer file cabinet next to the desk and the files kept in there locked. Terminated employees can go out to long term storage after the year ends and/or Cobra is finish.

Filing - Repair orders

If your DMS stores a PDF copy of the repair order, then you might only want to have a copy filed by the service department and the answer to who should file is easy – the service department! In fact, they really don't have to file, instead have them label a recycled box each month with the start date and let them just drop them in. When the box is full, enter the ending date and store upstairs. If they need a copy later, then they can find the box and yes, they might be slightly out of order, but it sure beats filing. When your storage time is up, then destroy the box. You no longer need an accounting copy if you can view the repair order online and run a report to reconcile cash sales instead of using accounting copies. It really is easier!

Filing - RO hard copies

These are becoming a thing of the past as technicians write their notes directly into the Technician Terminal program. If you are still using hard copies for your repair orders, the typical paper is 32#. This premium paper costs four times what normal 20# paper does. If costs even more if you're getting the hard copies printed in a buff color with lines and spaces for notes and flag sheets. With electronic dispatch and technician terminals, the need for these heavy hard copies has been decreased and most can be eliminated – or at least downgraded to a nice inexpensive 18-20# paper. For some of you, the 20# paper is free due to your laser fee contract, but you're getting charged a ton for the premium paper. Either way, you'll also save on shipping costs.

Finalize deals on DMS - reconcile to sales gross

The biggest key to a quick close is having all the deals in the office. That is way car deals are the first thing that I suggest you "flag." I have freely admitted over the years that I used to spend more time thinking about closing the books than ways for the dealership to make more profit. It is a good thing to find out later that being proactive about getting all the deals posted does mean more profit. If you can't do my flagging idea, one of my favorite bonuses is to pay billers a bonus based on the average days between delivery date and posted in accounting date. This might seem unfair since they don't control the deal flow – but they can encourage the reduction of this time lag. That means that your billers need to be proactive about getting deals posted. Instead of waiting for the deal to come into the office, billers need to go into the F&I office each day and finalize deals. A lot of us were trained that we didn't post a deal into accounting until it was "done." That brings up the question of "when is a deal a deal?" I agree that we should hold commissions for missing paperwork, amounts due, and credit approval – but we shouldn't avoid finalizing a deal. My rule is "if the customer has the vehicle, the deal must be posted in accounting – no exceptions." Having the vehicle means that the customer is driving on a temporary report of sale and not on a demo plate. Let's create a better method to track your car deals. Step 1 – Control your report of sales. Your first step to getting this process under control is to determine how customers can leave with a vehicle. As more states start to control this function better, there is often an item called a "police book" which records the temporary sales forms. Your billers would take responsibility for each item in this sales record. If you don't have a police book, then create one. Find the temporary forms and give them to someone who can check them out with a log. If they are not numbered, then number them in a area that is okay with your state. Put the numbers in an Excel spreadsheet and require an accounting for each report of sale. Step 2 – Post the deals early in F&I. For each vehicle in the in the new log with a new report of sale number, finalize the deal into accounting. Many times there is missing information like the buy rate for the financing, but that can all be done with a deal adjustment posting. The key items to post are the sale amount and vehicle receivable/contract in transit. Every deal will probably need a cost adjustment, finance reserve posting and commission accrual but that can be done when you complete the deal. Having the deal posted into accounting means that our DOC is more accurate and we know where our money is. In addition, if you have a floor inspection, you won't be scrambling for missing vehicles. You don't need to do a complete rebill of the deal. Instead you'll make a few entries into the same journal using the same reference number (or if your system doesn't allow that, enter an R in front or behind the reference number.) This early posting of the deal normally will happen every morning in the F&I office by your billers. It will be hard to train them – but by you watching that log in step one, you'll know every day if a deal snuck by and didn't get posted today. It's a new process so it might take months to get your biller to adapt to no longer waiting for the deal to come to her. That is why a bonus helps. Step 3 – Complete, title, commission and file the deal. This final step is done after all the money is collected. This means that your F&I managers need to pull the contracts and submit them directly from their office. Now that all the money or contract has been paid, the biller can tear the deal apart, title the vehicle, pay commissions, and make the final adjustments to the deal. This system is more of a "touch it once" concept. We seem to handle deals over and over – waiting for the money to get paid, holding the commissions, making adjustments. Now When is a Deal a Deal? With this process, your biller won't work the deal until it is all done and no more rush to get it into accounting because you posted it days ago down in the F&I office. How does this three step process create more profit? First of all you're going to pull that contract days before you're doing it now. Next, you'll have better control of where your funds are held up. Finally, you can pay off the floor plan earlier and that is a huge savings. In addition, this early posting brings your trades into stock earlier so you'll sell them sooner. Sure you're going to have a few more "unwinds" in accounting, but the other benefits outweigh the amount of extra accounting that has to be done – and that's what we're there for anyway!

Insurance claims – accident reports

Accident reports should be done promptly while the facts are still fresh in everyone's head. The best process is for everyone to first email the person in charge of accident reporting. I like this to be a chore for the payroll clerk or office manager. That person would then fill out the accident report based on the email and email any questions back to the person reporting the claim. Make sure that the person gets a copy of the police report emailed also emails an estimate for repairs.

Internals for PDI – New Open

You might think that adding an internal for a new vehicle should be the job of an advisor, but I don't believe that you should interrupt the work of someone who should sell for a job that is already sold. Every time you receive a shipment of new vehicles, have the clerk who stocks them in also open up a PDI internal to enable the job to quickly get dispatched without waiting for a dispatcher. Teach the clerk the correct OP code to use and it is done!

Internals for Used Purchases – Open

If every used vehicle needs a safety check or other mandatory work, then every time you get a shipment of auction vehicles or buy a used vehicle, then open up an internal for the vehicle. Put in the notes that approval for all other work must be approved and teach the clerk the correct OP code for the safety check or other required work.

Internals for Used Trades – Open

If every used vehicle needs a safety check or other mandatory work, then every time you work a car deal with a trade, then open up an internal for the vehicle. Put in the notes that approval for all other work must be approved and teach the clerk the correct OP code for the safety check or other required work. Put this step on your car deal close list.

IT – Add new users to DMS

Consider adding a new user with the least amount of security that you can and increase it as they learn the features. This is possible with most DMS and CRM systems that have dynamic security profiles. A level of a salesperson can manage their leads, create appointments and track actions. When the salesperson has learned that area and is ready to do more, then we upgrade their security to a CRM salesperson plus. That level can do some data mining and send out emails and letters. If the dealership wants them to quote payments, then we change their security profile to CRM salesperson plus with Quotes. The first step is to find out if you can customize your security and create profiles for your users. Next, ask a user that is having difficulty exactly which features they need today to do their basic job functions. If their only function is entering leads, preparing checks, or closing repair orders then provide access only to that function. If you're training a new employee, give them access to what you're training them on today, and then add access as they take a class on a new feature. One of the best utilized DMS systems out there requires regular testing for any function that the user has security access for. I'm not suggesting that, but it does prove that training does work. Instead of creating more work for your employees, why not try to simplify their day by removing some of their security access? It will be a treat for them to get the function later – like a free App, when they are ready to learn more.

IT - Add users to the network, passwords

When you add users to both the network and DMS consider using their real name – like Sandi Jerome instead of "hotlady37." This makes identifying the user easier and a more professional look on the PC when it shows logged in user. Let the user be creative in their passwords, but the user name should clearly identify the person.

IT - DMS system administration – backup

With newer DMS systems no longer need to do a daily backup but if you're still required by your older system, then this is a process that should be done by the system administrator before they go home. If possible try to get it automated. The first thing in the morning, take out the backup tape (if you use that) and put it in the fire proof safe and put the next rotation in the drive so that it is ready for that night backup.

IT - Email system

I'm not so sure it is necessary for dealerships to have their own email server. Most website providers also provide mailboxes and a method for the employees to manage their email – without Outlook. Why not look at Network Solutions or even Gmail from Google. Over the past year, I've been trying to do more with the free Google tools. I quit paying for AOL (although I've kept my AOL email address) and added a Gmail account. The only problem is now I have more than one email account; yahoo for my cell phone and two Gmail accounts. If you use Gmail for one of your e-mail accounts... did you know you can add all your other email accounts to Gmail and access them all from gmail.com? Just go into your Gmail Account Settings and click on Accounts to add them. http://www.gmail.com. Now all my emails; sandi@SandraJerome.com, sandi@sandijerome.com, etc – all go to Gmail where I can filter, delete and then decide what I want to download to Outlook later.

IT - Internet access approve and review

It is pretty hard today to NOT provide a user with Internet access. It is important to restrict the sites they may visit. Is your Internet slowing down? Are you concerned that your employees are downloading pornography – or spending too much time on gaming or video sites? There are many companies that provide web blocking software; DynaComm, i:filter,iPrism, SurfControl, Websense, and X-Stop R2000. I'm most familiar with Websense, but most have similar features. What does web blocking do? First of all it detects security risks or productivity problems arising from employee Internet activity. Next you can manage employee Internet access by using a database that Websense maintains – or employees can request access to the sites that they need and then you approve the request – or deny it. Best of all – you can block spyware, phishing sites, and malicious mobile code. Worried about getting sued for file sharing? It does P2P blocking and manages instant messaging and IM attachments. To keep your Internet running fast, you can manage the use of streaming media and other high bandwidth applications. Finally, it prevents employee hacking by restricting access to files and hacking tools. It's an expensive product – but it might pay for itself if it keeps you from being hacked or sued. Your first step is to make a company policy that Internet use is strictly for business use and not for browsing YouTube, Facebook, or being in Chat rooms.

IT - Virus protection

This is another point for trying to switch from network servers to more "Cloud" computing. Having to protect a server from attacks is tedious but extremely important. I know one IT director that spent the night in the dealership trying to catch an attack on his server. Unfortunately, most of these attacks are done by programs and not humans. It's not personal, they are just out there. Make sure on a daily basis that you're not under attack and that all PCs that virus protection, and that the server's virus protection is updated.

Lien payoffs – Shortages

As soon as a payoff is made and the estimated payoff is short, make sure you email the salesperson, manager, and F&I manager. It helps to also write a quick letter to the customer unless one of the others will be doing this. Quickly look up in service and see if the customer has a first service appointment scheduled and put that on the company-wide calendar and a note in the appointment file.

License fees – Shortages

As soon as a title/registration check is issued and the estimated license fees are short, make sure you email the salesperson, manager, and F&I manager. It helps to also write a quick letter to the customer unless one of the others will be doing this. Quickly look up in service and see if the customer has a first service appointment scheduled and put that on the company-wide calendar and a note in the appointment file.

Lien Payoff trades-in - pay

It is important do these payoffs daily – as soon as the deal is final. Run a report of all pending deals and find any trades. Sometimes a deal gets held up in F&I for an unrelated issues like waiting for a rebate or title form and then the payoff gets out of date. Create a report that has the payoff date, deal status, and other information and check it each day for upcoming payoffs. Make sure you have another field in the file where you can enter the date the payoff was made so you know if payoff have been made.

Mail – distribute

The best mail system is a collection of mailboxes where employees can retrieve their mail daily. Why should this be done by the receptionist? Do you think that the new head of 3M opens the mail? In most companies, opening and distributing the mail is done by the lowest paid employees; the mail room. Consider having your receptionist open the mail and hand it out. Do you need to monitor checks received or complaint letters? Fine, create a report for her to complete. Want to make sure nothing odd is happening? Great, then look through the mail once a month – or once a week just to keep everyone on their toes.

Mail - outgoing to post office

Even though you might have most of your mail going to a Post Office Box, try to keep your street address for mail delivery and get your magazines and catalogs sent there. It helps cut down on the amount of stuff your employees must carry. The best person to take out the mail is the parts driver who is always on the road anyway. This person can also pick up the mail at the post office. If you can still get street delivery, there is another backup for making sure things get in the mail – the same carrier that delivers can also take the mail – along with the carrier who might be stopping by to pick up packages in your parts department.

Net Profit Estimate

This daily report is much easier for a dealer or GM to understand than a DOC. It prorates out the deals, parts and service gross and estimates that based on how many days are left in the month. It uses a variable rate for variable expenses and then a fixed amount for other expenses and a line for the extra income from dealer cash programs.

OSHA claims – Submit and Record

Within eight (8) hours after the death of any employee from a work-related incident or the in-patient hospitalization of three or more employees as a result of a work-related incident, you must orally report the fatality/multiple hospitalization by telephone or in person to the Area Office of the Occupational Safety and Health Administration (OSHA), U.S. Department of Labor, that is nearest to the site of the incident. You may also use the OSHA toll-free central telephone number, 1-800-321-OSHA (1-800-321-6742). Recording; You must consider an injury or illness to meet the general recording criteria, and therefore to be recordable, if it results in any of the following: death, days away from work, restricted work or transfer to another job, medical treatment beyond first aid, or loss of consciousness. You must also consider a case to meet the general recording criteria if it involves a significant injury or illness diagnosed by a physician or other licensed health care professional, even if it does not result in death, days away from work, restricted work or job transfer, medical treatment beyond first aid, or loss of consciousness.

Parts Inventory Aging Analysis

Monthly enter the balance of the parts inventory and the percent that is overaged. You can also detail out the top oldest parts, parts with the greatest value, etc. The key is to look at the reports available in the parts reporting area and trend these to monitor your parts inventory.

Payoff - Floorplan/EFT

I like using the accounts payable program to handle floorplan payoffs. Create an AP vendor for your floorplan only. When the biller works a deal, she debits the floorplan account and credits the AP account using the deal, VIN, customer for the reference. Daily, before cutoff, bring up the AP account and payoff all the floorplan. If you need the biller to provide more info like the lender, sales date, etc.; just get that setup in your interface.

Payroll - garnishments, setup

As soon as you receive a garnishment, set that up in payroll – on a daily basis. Don't wait until you're trying to run a payroll – otherwise it might get missed and then you're liable for the amount.

Post EFTs for contracts paid

If you are using the AP method then you don't need to post the EFT for the floorplan. Just print the "check" on blank paper – or save as a PDF document.

Prepare key tags for New

I'm a firm believer that the key tags need to be prepared BEFORE the vehicle arrives on your lot. This means getting transportation notices faxed or emailed to you. Then prepare the key tags and leave them with the person who stocks in vehicles so that a key is never left "untagged." If you use the VIN# for the stock# then it is easy to match the key tag to the vehicle arriving.

Prepare key tags for Used

As with new, key tags need to be prepared BEFORE the vehicle arrives on your lot. This means getting transportation notices faxed or emailed to you from the auctions or your buyers. Then prepare the key tags and leave them with the person who stocks in vehicles so that a key is never left "untagged." If you use the VIN# for the stock# then it is easy to match the key tag to the vehicle arriving.

Receipt in Down Payments

Almost anyone in the accounting office, receptionists, cashiers and F&I managers should be able to receipt in a down payment. When someone finds a check in the deal – receipt it! No customer should have to wait for someone to prepare a receipt. If everyone knows how to find a customer number on the DMS – then adding a customer down payment is easy. There should always be a deal started – no exceptions! How can you take a down payment and there not be a deal yet? Learn how to search for the customer number in F&I and there will be no need for accounting to add a customer/number. Have everyone put the receipt number on the check and put into the night drop if it is easy to find – or use a central cash drawer. Down payments ususally do not need change – so no need for that fire proof cash drawer to have a "fund".

Receptionist - relieve for breaks

Nobody in the office likes to relieve for breaks, but if you rotate the task each week between 2 members of your office staff, then acceptance is a little easier. The best person to do the relief is someone who works with the sales staff – like the person who works the heat sheet or stocks in vehicles. This gives the person an opportunity to ask questions they might have while they are "out there." This person would then transfer that chore to the other person on your staff that did the relief last week. Try doing it yourself too – sometimes a breath of fresh air is good!

Purchase orders – issue for Expenses

There are 4 types of purchase orders; parts, sublet, vehicle adds and other. The "other" are for expenses like office supplies, advertising, etc. should be done in the office and by the AP clerk since that person can also convert the purchase order into an invoice if they already have an invoice. All purchase orders today should be on the DMS and this means training a lot of people on how to properly issue a purchase order.

Purchase orders – issue for Parts

There are 4 types of purchase orders; parts, sublet, vehicle adds and other. For parts, obviously these are done in the parts department. All purchase orders today should be on the DMS and this means training a lot of people on how to properly issue a purchase order.

Purchase orders – issue for Sublet

There are 4 types of purchase orders; parts, sublet, vehicle adds and other. For parts, obviously these are done in the parts department. Sublet is often done in parts, but I prefer that service advisors issue these purchase orders since they are the once that have to collect the money on a repair order. All purchase orders today should be on the DMS and this means training a lot of people on how to properly issue a purchase order.

Purchase orders – issue for Vehicle Adds

There are 4 types of purchase orders; parts, sublet, vehicle adds and other. For parts, obviously these are done in the parts department. Adds to vehicles should be done by the sales managers or inventory manager. All purchase orders today should be on the DMS and this means training a lot of people on how to properly issue a purchase order.

Report new deliveries to factory (RDR)

This is a task that is best done in the sales department, by the F&I manager if it is important that deliveries be reported promptly. Often reporting the delivery is combined with submitting for rebates and having the person who know what rebates were used in the deal is very helpful.

Rebates - Submit

Same as reporting deliveries, this is a task that is best done in the sales department, by the F&I manager who know what rebates were used in the deal. When the biller gets the deal, they can check the submission form to make sure the rebates were submitted and apply for any that were missed.

Repair Orders - open over 3 days old

Depending on your manufacturer, adjust the "3 days" to a reasonable amount. For example if your factory is Audi and it takes 4 days to get most parts, then you'd want to adjust to 5 days. On the other hand, if you're a Ford dealer with a depot and 10 Ford dealers nearby, most vehicles should be fixed daily - so 2 days is more reasonable. Run the open repair order from your DMS; for ADP the job is MR RWP, and Reynolds + Reynolds ERA 3613. It is also important to run a report of those repair orders that are closed in service and are waiting to be updated to accounting. They are: for ADP, DR RAC, and Reynolds + Reynolds 3614. These lists will indicate which repair orders are still being worked on or have been completed but not paid for. Many on the list could be for vehicles that have been taken by the customers without payments. It is important that an explanation be provided for each item on the list.

Sales commissions - calculate, accrue/voucher

In my monthly section, I discuss eliminating this task with a report, but if your commissions are too complicated, then sales commission should be part of the car deal process and hopefully part of the DMS recap or commission logic. See if your DMS system can create the voucher and then during the posting credit a commission payable account scheduled by the salesperson's number. Have another control number to hold commissions. For example 123 would be the control number for salesperson #123, and 123H would be the control for the held commissions. When commissions are released, credit them to 123 and then at payroll time, pay the amount in the schedule and print out the schedule to attach to the checks.

Scan Car Deals

Some of the larger DMS providers have scanning systems, but car deal scanning is usually not included in their basic scanning option. To scan car deals, purchase an "off the shelf" scanning system that comes with an auto-feeder and great software. I use an inexpensive HP scanner and Windows Scan (free) but if you're going to do this right, then spend your money in the search/index features of the software and speed of the feeder.

Scan Parts Tickets

Yes, some of the larger DMS systems have document storage, but this is usually proprietary and you lose this data when you change systems. If you want to test this, ask your DMS system how you'd get all the parts ticket out for a certain year to a CD or USB drive for an audit, and you'll find out that you can't – without bringing each one up at a time and saving to a PDF file! To do your own document storage, rather than printing the accounting copy, save it to a PDF file instead. If all the information about payment is in the document, then why write that info on a copy and scan it. Many of the newest DMS systems allow you to save it to a PDF file, or you can add a free PDF printer (I use BullDog) and print the accounting copy to the PDF printer and save with the parts ticket number. To save a copy for the customer, save as 12345C and 12345A for accounting.

Scan Repair Orders

Yes, some of the larger DMS system have document storage, but this is usually proprietary and you lose this data when you change systems. If you want to test this, ask your DMS system how you'd get all the repair orders out for a certain year to a CD or USB drive for an audit, and you'll find out that you can't – without bringing each one up at a time and saving to a PDF file! To do your own document storage, rather than printing the accounting copy, save it to a PDF file instead. If all the information about payment is in the document, then why write that info on a copy and scan it. Many of the newest DMS systems allow you to save it to a PDF file, or you can add a free PDF printer (I use BullDog) and print the accounting copy to the PDF printer and save with the repair order number. If you have multiple copies then save as 12345W and 12345C for Warranty and Customer copies.

Scanning other Documents; Checks, Receipts, Purchase Orders

With most DMS systems, you won't need to scan these since a type of mirror copy of the data should be in the system. Check how long data is saved to make sure and if all the information is easy to find; payee, person who prepared, etc.

Send contracts to the bank

The best way get your money quicker is to have the F&I managers pull contracts. I used to pull my own contracts when I was an F&I manager because I knew exactly what was supposed to go with the contract. It was easier to do it myself then write a note to the office telling them what the bank wanted. My dealer would come into my office each morning and sign the contracts and then I'd take them back to the parts department to have them deliver or overnight. Do you think your F&I doesn't have the time? You'd rather have your F&I concentrate on getting deals done or selling products? Do the math. An F&I manager can do 100 deals a month – that works out to 4-5 a day, but because the weekend is so busy, the weekdays often go by with only one or two deals – ALL DAY! They have the time pull contracts – it only takes minutes and they know when it needs to go in. How much can you save? For a typical dealership, 2 days can add up to $2349 a month in savings! That is almost $30,000 a year! Can't convince
this F&I manager? Just wait until you hire a new one and train that one to pull contracts. In my seminars, about 20% of the attendees are having F&I managers pull contracts with the dealer signing them in the F&I office.

Service Vehicles Not Ready When Promised report

Two things can make this report a problem; the first is when vehicles are really not ready when promised and the 2nd is when the wrong (default) promised time is entered for each repair order. If all the promised times on the report are the same, then you probably have this problem. This report should be run each morning for the prior day and given to the service manager. You can also run it in the afternoon at 4pm for vehicles that day.

Service hours Left Unsold in the Shop

I don't know a dealer in this country that doesn't ask this question every morning, "How many did we get out yesterday?" On the flip side, I only know a handful of dealers that know how many hours the shop did yesterday. It is not a metric that we track – but imagine if when you asked your sales manager how many he sold yesterday he responded, "we got ten out, but all the rest of the inventory is gone this morning." That is what happens in your service department. Every hour that wasn't sold yesterday is gone today. You never get it back. How can you create the kind of report that helps you monitor this metric daily – even hourly? A flash report is the perfect solution. What is a flash report? It is a snapshot of the situation for a particular department or profit center with the objective to make the changes you need to increase profit.

Service hours sold in shop yesterday

Who should get this report daily? Give it to the service advisors. Humans are naturally competitive and neither Sam nor Sue is going to like seeing those big negative numbers every day. They will get proactive to making the changes that they need. How do you create this report? It is not easy. First of all, a flash report means that it is a "snapshot" of your situation. To get a good snapshot of anything – you need to get the subjects together in the same picture. For the hours sold report you'll need three reports from your DMS system; a report of appointments, a report of carry over or open repair orders that are ready to go, and a performance report that shows hours sold month to date. You can download these reports and put them on a separate tab in this spreadsheet and then use either VLookup to fill in the cells on the report or just link to the cells. To make the report accurate you'll need to make sure that you're using the appointment system and entering estimated hours correctly. You'll also need to look at your dispatching system and determine if your repair order status is correct. Finally, are your booked hours accurate and up to date? Don't hold open completed repair orders for any reason. If the vehicle is gone and the repairs are completed, then close the repair order. If you've ordered a part and let the vehicle go, then move that part to your SPO tracking system and open a new repair order is you must. Do not keep warranty repair orders open, close and transmit daily. If you can't do this, then you have a problem with your warranty submission process. It is usually a rule of thumb that the warranty clerk with the lowest reject rate holds them open for the longest period. Do you have a problem? Run a report and compare booked date to closed and transmitted date.

Sign Checks and Contracts

Getting interrupted throughout the day to sign a check or contract is frustrating. Make a time each day when you'll sign them and be faithful to that time. 3pm is a good time and put the bucket outside your office on the file cabinet. Sign them and put them back there by 4pm.

Super Numbers Report - Prepare

"Super Numbers" is what I call a DOC (Daily Operating Control) for the Controller, Office Manager, or CFO. It's a set of key indicators as to the health of the dealership operations. Before you can create your Super Numbers, you need your schedules and balance sheet items to be correct – and then create a system of reporting that provides this information. As a tip, I recommend that Contracts in Transit should be EFT contracts paid in 2-3 days – only put in this account money that you will be getting quickly. Vehicle receivables are special financing, returned contracts, one-pays — anything that might take longer. If you only have one account for both of these items, create a sub account for items that take longer than 3 days to be paid. In your AR and Vehicle receivables you can't have frozen amounts like intercompany receivables or notes from the dealer or employees. These accounts need to hold amounts that you expect to receive within 30 days. Same rule applies to your used vehicle inventory. That's not a place for the parts truck or employee demos. Make sure that you can send those vehicles to the auction if you get in a pinch. In addition, your cash balance has to be real – do not hold checks – or if you do – add them back into your Super Numbers Cash amount. The purpose of knowing these five Super Numbers each day is to know where you stand and where you can drum up cash if today is the day the bank wants their million and a half.

Supervise cashiers - hours worked - overtime

I like the supervision of the cashier to go to someone in the office that is being trained to be an office manager. Since this can be a time-consuming task, delegation is important. A good person is the same person that totals out the daily deposit for parts and service since they already have interaction daily with the cashiers. They would also schedule the relief for the cashiers. Make sure this person reviews the hours worked each day and watches for any potential overtime problems. Another option is to give this task to the warranty clerk.

Supervise office staff hours – overtime

Make sure your office staff knows that any potential overtime must be pre-approved. Normally you should not have to supervise their hours – they should be able to self-supervise if they know that overtime is an issue. They must report to you at 3pm if there is any task that must be done today that it doesn't look like they will get to within their normal hours so you can get them help. Hopefully your DMS has an overtime watch report that you can look at during the day to find any potential overtime problems.

Supervise phones, receptionist, hours worked – overtime

I like the supervision of the receptionist and phones to go the sales managers since they are ones who notice the performance of this person the most. Since this can be a time-consuming task, "letting go" of this control is important. I don't know where it became that a greeter is supervised by the sales manager, but a receptionist by the controller? If possible the sales manager could also schedule the relief for the receptionist and it could even be a salesperson. Make sure the sales manager watches for any potential overtime problems.

Take Deposit to the Bank

Find out the cost of an armored car service. When one of my clients asked their bank, they offered to provide the service for free! No more trying to find a demo - or someone to take the deposit to the bank. Another option is to ask the dealer to take it daily. They often enjoy the visit. Doesn't hurt to ask. You might see if the parts runner can do this – but make sure they go straight to the bank. Leaving the deposit in the parts truck while delivering parts can be risky – same if you have a valet service, make sure it is only done when they don't have a customer in the vehicle to prevent theft.

Titling of vehicles – New

Today, many states provide online titling or even enable the dealers to issue their own plates. It is more important than ever that before a vehicle title is transferred that you verify that funds have been collected and the correct leinholder is on the application before processing title applications. This means that a process needs to be in place that when a payment is received from a leinholder, a copy of the check is given to the title clerk for verification of the title application.

Titling of vehicles – Used

As with new titling, many states provide online titling or even enable the dealers to issue their tabs. It is more important than ever that before a vehicle title is transferred that you verify that funds have been collected and the correct leinholder is on the application before processing title applications. This means that a process needs to be in place that when a payment is received from a leinholder, a copy of the check is given to the title clerk for verification of the title application. If the vehicle had a previous leinholder, then a process that pays off the trade-ins promptly will make the titling into the next owner's name easier.

Wholesales deals – post

Most dealerships have a "biller" or new and used car desk that posts retail deals or you might be adopting my Tasks process. Wholesale deals are much different – actually very easy and can be posted by another clerk – often the inventory clerk or title clerk is a good person in a larger office. You can even train a new person to do this as they are learning other office functions. Accounts receivable is another option since they might get the job of collecting form Wholesalers.

Wholesales get paperwork ready for auction/wholesalers

The very busy new and used billers are typically interrupted when vehicles need to go to the auction or wholesalers arrive for paperwork and titles. Again, the inventory or title clerk is a good person to deal with this "on demand" work – or the same person who does your dealer trades. When I worked in a large office, I was the contract desk. All deals flowed from me in the morning where I made sure the contracts had been pulled and submitted by F&I and monies were collected. In the afternoon, I did the wholesale deals and dealer trades. I wasn't as qualified as the very seasoned billers, but as a newer office person I learned about deals before moving into the more qualified (and better paid) biller positions.

Weekly Tasks

Why perform a task weekly? Weekly tasks normally can help you make more money and normally analyze items like technician productivity, closing ratios, and labor gross profit. They also are involved in collections – helping you keep the money you have. This is the time to keep schedules clean instead of monthly when you're trying to close the books.

AP - Schedule clean

If you download the schedule to Excel and filter on any debits and investigate. Next filter on small debits. Are they mis-postings, less than $1.00? As a whole, there should be no debits on this schedule, so keeping it clean weekly will save you time during monthly statement reconciliation.

AR - NSF check receivable – Review

Putting the responsibility to collect the NSF check into the hands of the employees who know the customer and are paid on the profit from the sale helps the collection process. To report on all NSF checks, make sure that sales NSF checks go on the daily Heat Sheet and service/parts NSF are on the AR schedule. Make 3 copies of the NSF check or scan into your computer. Email or give one copy to the sales person, counterperson, or service advisor. The 2nd copy goes to the department manager, and the 3rd copy to the general manager.

AR - Past Due Report to Parts and Service Managers

The best past due AR report comes from your DMS system. Often your AR schedule will work. Make sure that it has these fields; Customer name, contact name, phone numbers, Last paid amount and when, Last charge amount and when. Email this weekly to enable the managers to collect when they see the customers. Don't be afraid to ask your managers to "hand deliver" a difficult AR account's statement.

AR –Review with AR clerk

Collecting body shop receivables is much different than parts - and service is also different. It helps to review them by type and I actually like Body Shop receivables to go onto their own schedule, similar to Extended Warranty claims. Your solution for a past due parts ticket might be reviewing the last payment stub and seeing if the customer needed a copy faxed. If this is a happening often, then you can review the PROCESS instead of the situation. Why are customers always missing their copy? If possible, separate the receivables into parts, service and body shop receivables for management review. This can be done by using a separate GL account for each type of account or use a special "type" field. Another method is to assign customer numbers with a prefix for the type of customer. If you are using the phone number for the customer number, prefix it with a "P" for parts accounts, an "S" for service accounts, and "B" for body shop accounts. Since the schedule prints in control number order, the B's will print first and you work on those first. What is the process by each insurance company? Who is in charge of making sure the insurance company has what they need.

AR – Past Due Meeting

There needs to be a weekly Fixed Operation Heat Sheet that contains;1. Past due AR account report 2. Warranty claims over 30 days old or rejected 3. List of open repair orders with labor and/or parts charged on the RO 4. Special order parts not picked up

AR - Schedule clean

If you download the schedule to Excel and filter on any credits and investigate. Next filter on small debits. Are they mis-postings, less than $1.00? If you have multiple schedules, download together and look for errors on the account, control. Be sure and correct the process – why don't the clerks know the correct account for body shop payments vs. parts customers? Create a cheat sheet if you need to.

Comeback Report

Create a report of all the comebacks and how it was handled. When a customer brings the vehicle back for the same problem, this is called a "comeback repair". Service departments handle comebacks in different ways. Some service managers require that the technician that fixed the vehicle originally fix it again for free. Others will assign a new technician to the job and "back flag" the original technician for the time. The most common way is to assign a new technician to the job and expense the cost to policy adjustments. The reasoning behind this is to not spend a lot of wasted time finding blame, but rather to get the customer's vehicle fixed by the best technician and have the expense paid by the service department. If a certain technician must have a large amount of his repairs redone, then he is either coached, trained better, provided the tools he or she needs or is terminated. Hopefully you already have a report that you run to track how many comebacks were caused by each technician. Some service managers track this "in their head", but there are better methods. Some of the computer systems have a comeback report by technician, but often it incorrectly gives the comeback repairs by the correcting technician also. Understand how it is setup before you start providing this report - hopefully every week. Another issue might be your pay plans. If advisors are paid on service sales - and comebacks go to an expense account, then it would be human nature to write as many comeback repair orders as possible. This makes technicians happy - getting to redo someone else's work and advisors make the commission on the easy sale to an expense account. I prefer to not pay advisors on comebacks. All comebacks must be written by the service manager (with his or her advisor number) and thus they don't get paid to advisors. Comebacks should be done with coaching and the help of the shop foreman by the same technician with no flat rate hours - so the expense stays low. With the service manager handling the repair order, hopefully they will get to know the issues causing comebacks better and more importantly, make sure this customer has their vehicle fixed right the 2nd time.

Deals That Accounting Gross Does Not Match Sales

To create a report of all deals where the accounting gross does not equal sales gross requires a two-step process. First, you need to get your accounting gross profit. Some DMS systems store this in the journal file (R+R) as a field that you can query, but others like ADP, only calculate and print on the journal. To get this from the journal, you'll need to print the journal to a print file and then get the profit line and stock number on the same line. Next, query your deal file, for example; FI-WIP for ADP, FIMAST for RR. Get the stock# and gross profit. Bring these two spreadsheets into one workbook in Excel and use the Vlookup to compare the gross to each other.

Deals That Gross Profit Is Less than Benchmark

I like to make sure I get a final deal report from my DMS system each month. It should have the buyer information, salesperson, bank, vehicle description, and profit amounts; gross profit, extended warranty gross, etc. Since your gross profit benchmark can change each month, send an email to your general manager or deal and ask for the current benchmark for new and used. Then bring up this report and use your Excel autofilter to find deals where the gross is < the benchmark amount. Before sending to your general manager or dealer, study the deals. For example are used vehicles that are trade in in the list? Why would new car trades have a lower gross profit than purchased vehicles or used car trades? Remember – don't give management the "what" of "new car trade gross profit is low" – give them the why… Who appraised the new car trades with a lower gross profit? Who buys the new car trade-in? Are they retailed? How many days does it taje? Were the new car trades wholesaled and to just one buyer? Your new car trades could also have a higher gross profit – this happens when the used car department "steals" the new car trades and causes an understatement of new car gross and over payment of used car commissions or even covers some used vehicles wholesale loss.

Employee receivables - Schedule clean

I don't believe employees should be able to charge their repairs or parts tickets to an account – no more than a regular retail customer. Encourage employees to use their credit cards to avoid the hassle of having to take it back out of their paycheck. The labor involved and the write off of old employee receivables (that got missed before they left) far exceeds any discount fees. The employees will also earn miles that they will enjoy. Unfortunately, garnishments and insurance premiums mean that we still have an employee receivable account. This should be reconciled the day before payroll – if you're doing weekly payroll, then weekly – otherwise twice a month for semi-monthly payroll.

Floorplan - Schedule clean

This is different than the monthly reconciliation to the lender statement that you'll find in your monthly chore. The purpose of a weekly floorplan reconciliation is to quickly catch any double payments, payment amount errors, or an error that comes from swamping floorplan in dealer trades. First download the schedule to Excel and filter on any debits and investigate. Next filter on small credits. Next, download inventory/floorplan together and subtract the two. Filter on any large difference to find unfloored units or other problems.

Holdback - Schedule clean

The holdback schedule is one of the easiest schedules and should be "cleaned" every time a new credit memo is applied – which is normally weekly. This can be one of the first schedules that you assign an "A" Skill – Anyone person. The biggest problem comes from "swamping" holdback during dealer trades. Carefully study how you are paid holdback (based on the selling or invoiced dealer) and the best way to make sure during a dealer trade that the correct entries are made.

Journal Gross Review

Go through your journals and review the gross profit percentage for labor and parts looking for labor less than 70% and parts less than 35% on repair orders and 25% on wholesale parts. You can set these percentages to better fit you dealership. In new and used, do a sales gross vs. accounting gross report – but still view the gross for any units sold at a loss or a huge profit, to get posting errors.

License fees – Schedule clean

This schedule should be reconciled each week, hopefully right before a heat sheet or manager meeting. Try to make it easy for customers to pay their license fee shortages - create an easy form letter in Word that enables them to send back with a check or fax/email credit card information.

Lien payoffs - Schedule clean

This schedule should be reconciled each week, hopefully right before a heat sheet or manager meeting. Try to make it easy for customers to pay their lien payoff shortages - create an easy form letter in Word that enables them to send back with a check or fax/email credit card information. Be sure and explain how the shortage occurred and a copy of the agreement that they signed to be responsible for any shortages.

New Inventory - Schedule clean

If you download the schedule to Excel and filter on any debits and investigate. Next filter on small credits. Next, download inventory/floorplan together and subtract the two. Filter on any large difference to find unfloored units or other problems.

New Vehicle Worksheet, Sold YTD, In Stock by Model

There should be a written policy used by the dealership to order new-vehicle inventory. Either the dealer or general manager should be responsible for making the vehicle ordering decisions and these decisions should be based upon an inventory worksheet provided by the sales manager. The accounting office can help produce the information that is needed to make the decision. The attached worksheet is a guide to ordering vehicles. It is important to know: 1. What has sold in the past by model or option. 2. What is in stock by model. 3. What is already on order or in-transit by model.

Parts override report

Overrides in parts need to be monitored. If a parts person wants to steal a part, the easiest way is to create an ticket with no sale amount. This prevents it from showing up on the plus/minus report that is printed monthly. Overrides are also time-consuming. Every time a part has an override, the counterperson had to make a manual entry. Try to get the pricing matrixes set up properly and you'll reduce overrides and increase productivity. RR-ERA, From the Parts Report Menu select job 2542. Choose the date range for override details such as employee, invoice number, part number and description, quantity, cost, original price an the override details ADP - ROV: Report Overrides. This report prints all overrides entered from OE (sales) functions. Setup prompting is line-by-line; it includes standard and self-explanatory prompts. As you set up this report, keep in mind that override records are created each time you change: COMP code, core charge, customer's credit limit, customer's number (to another customer number), customer's past due credit status, customer's taxability, customer's tax exempt number, invoice charge taxability, C.O.D. or cash sale to chg, sale type, tax amount, tax code, prices on a part sale (cost, list, sale or comp), bin location, part number description, part taxability, part sale account, refund of core amount taxability, wholesale compensation qualifying status of a part (prints W=YES or W=NO), service charge and service charge taxability. Override records are created if you change or add freight amounts and the freight type is not set up. Override records are created if you reverse a core sale and/or return that the system cannot fully verify. Example: If the detail records for the core sale invoice are no longer on file or if the sale and return records are not linked, the invoice numbers are considered overrides. The report includes. invoice number, customer name and number, date the invoice was opened, date the override was entered, part number if the override was for a specific part, override entered, customer's credit status, number of the employee who entered the override, and if applicable, core sale or return invoice numbers used on a core reverse.

Service Advisor Labor Hour Sales

This Service Advisor Analysis takes information from a standard Service Advisor report on a DMS system does more of a side-by-side comparison of advisors (most reports have separate pages for each advisor rather than putting them next to each other.) When you compare service advisors, there are a few things to consider; 1. What is the mix of customer and warranty to internals? If one advisor is doing a higher percentage of internals than the other, then find a way to balance that out. 2. How does there labor hours per RO and labor $ per RO compare? 3. What is the Effective Labor Rate for each advisor?

Service override report

For service, exceptions occur whenever you override tax amounts, change an operation code, make changes after printing pre-invoices, override a sale amount, or override promise dates or times in Dispatching. With ADP, Use function Report Exceptions (DR REX) to list all exceptions posted to repair orders since the last time you cleared the exception information. With R+R, Access the service exception report via menus main 3000, invoicing 3600, and reports 3610, and lastly exception report 3619. Choose the billing type, department, starting and ending date ranges, as well as make, model. You can also specify a port number or user ID. There is an extensive list of more than 40 exception types that can be chosen to be reported. I recommend that as a starter choose the sale and cost override exceptions. Some of my favorite exceptions are cost higher than sale, voided ROs and sublet markup overrides. Keep in mind that not only will this report help you to discover theft or improper discounts some overrides could be due to incorrect computer specifications, training issues. The Policy Adjustments/Service Discounts report lists all repair orders with policy work or discounts granted. Service managers can review this report to determine if policy adjustments, write-offs, or parts and labor discounts are being used inappropriately. The RO number, discount percentage, and name of the customer are all listed on the report, as well as the initials of the person(s) making the adjustment(s). To request this report, go to Program 269-12, Policy Adjust/Svc Discount Report, enter a date range and press transmit. If you wish, you can narrow your report to certain invoice types or pay types, or show only parts that have been discounted. Service managers can also review all ROs with forced pricing by requesting a Technician Hours Flagged report in Program 269-07 and a Work-In-Process Listing in Program 269-08. For the Technician Hours Flagged report, select a date range and the 'Forced Amount Only' option, and press transmit. For the Work-In-Process Listing, select the 'Labor Lines With $' option, enter a date range, and enter a 'Y' in the 'Forced Only' field. You can also narrow each of these reports to certain advisors, technicians, or pay types.

Special Order Parts Not Picked Up

A report of all unclaimed special order parts should be run weekly for the service advisors. They need lots of time to call the customer and arrange an appointment.

The parts department should also show the advisors how to do an inquiry on a special order part so they can help the customer know if their part has come in, without transferring the call to the parts department. This will help ease the phone load in the parts department and provide better customer service. Statically, approximately 80% of parts obsolescence comes from special order parts that are never picked up. Since the nature of these "special orders" will not generate enough demand to justify stocking the part, it remains in the inventory system as a "non-stock part". Another reason for having non-stock parts is through the receipting of "factory recommended" stocking lists. These lists are usually generated for new model vehicles based on demand for similar parts in the past. Typically, this type of "non-stock" part will achieve stocking status in a relatively short period of time. When a customer needs a part that the parts department does not stock, this becomes a special order parts request. The request is generated either by the retail counterperson, wholesale counterperson, or the service department. The parts department employee either adds the request to the computerized special order parts system or writes up a 2 or 3 part request form. The hard copy of this form is usually a postcard that can be mailed to the customer when the part arrives. Make sure you use your DMS system and not a “hand” system. It is important for two reasons; first it ensures that the part is tracked by the computer system and that the part is treated like a lost sale and counts towards phase-in criteria for the next stock order and second, in order to have a way to receive a report of the special order parts on hand. A lot of parts departments don't like using the computerized SPO system because it is easier just to complete a postcard order form and manually track the parts, but eventually many manufacturers will eliminate the annual or quarterly returns. Instead they will only allow the dealer to return a part within 30-60 days. With this new system it will be important to know which special order parts are close to being 30-60 days old so they can be returned for credit.

Service Technician Productivity, Opportunity and Efficiency

Productivity and Efficiency are the most common ratios that are monitored. Often the dealership mistakes one measurement for the other or uses total shop hours for available hours and actual hours. This is because it is difficult to track available hours and actual hours. Normally the first measurement, Technician Productivity will be enough of a guide to discover if there is a problem. Technician Productivity is; **Flat rate hours/ Available hours.**This is the standard for measuring a technician. Technician Efficiency is **Flat rate hours/ Actual hours.** Technician Opportunity is **Technician’s total available hours/Total hours shop was open.** If the shop is open 174 hours, but the technician was only there for 166 hours, then some opportunity for work to be performed was lost. This may not be entirely the fault of the technician; often training and scheduling could be the problem. The time card needs to be reviewed to find if the nature of the absences was due to: chronic illness, excessive training, lack of motivation, personal problems. Another measurement is Shop Technician Utilization; **Total actual hours/Total available**

hours If the technician only spends 150 hours on actual jobs, but was available for work for 166 hours, the low percentage would normally be a management problem. The technician was there and available to do work, but not enough work was assigned. The problem could be; failure of service advisors to sell available hours, poor dispatching, too many technicians, not enough business, poor follow up or promotions, time wasters. Computer reports; ADP - Report Technician Hours (RTH). Choose (N) to create a new report and then option five for technician summary report. Choose the time period for the information you are interested in and follow the prompts to generate the report. RR Select (3312) from the scheduling reports menu (3310). Choose technician recap report and select dispatcher, dept. type and output. This report gives "today," "week" and "month" calculations all on one page. For ARKONA users; To print the technician productivity report, take option 24=reports from the service department menu and take option 1=select by technician productivity report. Key in your date range and hit F4 to select one or more technicians, or just press [enter] to print for all technicians. The report is integrated into the attendance system to allow you to compare a technician's actual time on the clock rather than just a flat, eight-hour day. For UCS Power ; 1. From the power workstation service invoicing application, select print _ SDS/TEC/TTR reports technician productivity recap. This selection opens the reports (RPT) application. (You may also open the RPT program directly and enter SC in the job ID to access this report.) 2. Enter a date range for the report. Include the technicians you wish to appear on the report (or exclude technicians you do not want on the report). Enter a title for the report. 3. Specify any additional options for your report, such as whether you prefer a monthly or one line format and whether or not to include pay type or category code details and/or totals. You can also limit the report to certain category codes or choose to exclude certain category codes from your report. 4. Press F12 to request the report. 5. This report lists the number of hours a technician was scheduled to work and hours actually worked, productive and non-productive times, WIP, number of hours flagged and invoiced, and productivity, efficiency, and utilization percentages. You can also view totals for these categories by pay type.

Payroll – Tax Deposits

If you pay payroll taxes weekly, then there is normally a payroll tax deposit due each week. Check the dates and update your next due date and as an added measure send yourself an email reminder or set up in Outlook.

Payroll - Overtime report

Each week provide managers with a report of any employee that worked overtime. In your DMS system, you might be able to add a customized report to view daily and weekly overtime incurred all instances of employees who started work early or late, and all instances of employees who left work early or late.

Parts Fill Rate report

This report should be run every week at the same time as technician productivity. Often low productivity comes from the parts not being available and a measurement of what was filled from the shelf vs. same day vs. not filled is important. Consider paying the back parts counterperson on improvement of this metric.

Payroll calculate, update, print

This is in the weekly chores, but of course if you only pay employees semi-monthly or monthly, you'd adjust. This process is probably one of areas that can have the most improvement. If it is difficult to you to enter payroll – then it is time to find a way to make it easier. Excel spreadsheets can help you calculate overtime or union calculations. If pay plans are complicated, suggest ones that are easier to understand and still reach managements objectives to improve performance (consider my Pay Plan Wizard.)

Payroll, Technician time – submit

Regardless of when you pay technicians, normally the technician payroll or hours are reported each week for technician productivity reports. Hopefully you are not using the old technician manual flags. Send me an email if you want information on how to convert from the old manual (lick and stick) to electronic flagging (even if you don't have technician terminals.)

Payroll – hourly employee time submit

This task depends on when you pay your payroll. If it is weekly, then you need to run this report – or gather manual time cards at the end of the week. This is probably one of the biggest tips – find a way to get away from manual time cards and use a time clock system..

Post EFT for factory rebates and other credits

If these credits come in daily, then post them daily, otherwise weekly as a minimum. Waiting for the month end doesn't enable you to spot rebates that might have not been submitted or when they rebate amount was wrong and now you need to adjust commissions.

Postage - check unauthorized use

Once a week, look through the outgoing mail for any personal letters, bills that employees have added to the stack and make sure they have paid for the postage. Make it easy for them to pay for it – have a sign/chart next to postage meter with the rates, a small form to enter the amount and name. Have the clerk closest to the meter collect the funds. Sorry, no change – that goes into the coffee fund! Make sure that is on the sign.

Sublet - Schedule clean

This schedule could probably use a look every day by your cashier, but at least weekly you should run a report of this and compare to the work in process/open RO report using VLookup. If the amount is on the sublet schedule and the RO is no longer open – or a quick lookup of the RO shows no sublet entered, then quickly notify the advisor and ask them to open the RO or enter the amount. If you really have a problem, then pay the cashier a bonus for every sublet she finds daily that is not on a repair order. Go back even further and look for purchase orders issued to sublet that are not on repair orders. This is probably one of the biggest write off areas in the shop after warranty and policy.

Unapplied Time – Analyze Account

Each week, you should analyze and prepare a report on unapplied time. When the service manager or dispatcher decides to pay a technician for work that isn't charged out on a repair order, this is unapplied time. Some examples of unapplied time are: cleaning the shop or helping other technicians, paying a technician for more hours than the flat rate guide or warranty manual allows - usually done if the job was difficult or the proper tools were not available. Reconciling entry when adjusting the work in process account. (See tip on Work in Process.) The unapplied time account is usually a cost of sale account number that appears on the financial statement as "ADJ - Cost of Sales - Labor". If there is a large debit balance in this account, then you need to find out why. If the debits are coming from the payroll sources, review the daily time tickets to find the reason. There is a chance that training or sick time is being charged to this account. If so, suggest that they use the expense accounts for this purpose. If they are constantly paying the technicians for more than the guides allow, make a suggestion to management to minimize this practice. Any large debit adjustment from the general journal must be researched. See the comments in the Work In Process section later. If there are large credit balances in the account, then the service department is probably using an estimated cost for the technician time instead of actual. This practice should be discouraged since the actual cost is easily obtained. For example, if the dealership has 4 technicians making, $14 hour, $13 hour, $10 hour, and $8.00 hour respectfully, then the dealership may use $13 an hour for costing purposes. Since the actual cost is obviously going to be less, there will be a large credit balance at the end of each month. It is not that difficult to use the actual cost. The cashier merely looks up the rate for the technician number on the repair order and figures the cost correctly. If a dealership has computerized dispatching or ERO, there is no problem, the computer does actual costing if the technician setups are correct. If they are on manual repair orders, they often set up the computer journal to automatically cost based on either a flat rate or a percentage of labor. Although this may sound simple, it would be much more accurate to override the automatic costing with actual costs. Without this procedure, any of the other unapplied labor problems will be difficult to detect since the credits will overshadow the debits.

Used Vehicle overage report

If you have an aggressive used write down policy each month, then run this report at least a few days before the write down day – otherwise a weekly report after cleaning the used schedule is a good idea. Try to include the source of the used vehicle; new trade, used trade, auction, or purchased from…

Used Inventory - Schedule clean

If you download the schedule to Excel and filter on any debits and investigate. Next filter on small credits. Next, download inventory/floorplan together and subtract the two. Filter on any large difference to find unfloored units or other problems. If you are ending up with lots of debits and chargebacks, see my tip on closing internals faster.

Users on DMS - run list and verify

Each week, run a list of all the users on the DMS and verify that they are still employees or authorized 3rd party. This is why I like the user name to be the employees real name instead of "hotlady37."

Warranty Claims - Schedule clean

The warranty clerk should keep this schedule clean and apply the credits. Any write offs must be approved by the service manager and controller. Use the small balance write-off if possible. Download the schedule to Excel to find small amounts if you don't have this feature.

Warranty Claims - Schedule review

This schedule should be reviewed with the warranty clerk weekly, after all the credits have been applied. Look for processes that cause claims to be written off; parts not returned, work not authorized, vehicle didn't qualify, work done twice – and find ways to prevent this in the future. Determine if the technician or advisor should be charged back if it was their error.

Warranty Claims - Write off small balances

If you can set up these to be done on your DMS, then write off any amounts under $1.00, but find out how they are "getting off" because small amounts add up. If there is a set up error on the DMS, then get that fixed since you're either overpaying the technician, advisors, parts personnel – or all of them.

Warranty Credits – apply

If these credits come in daily, then post them daily, otherwise weekly as a minimum. Waiting for the month end doesn't enable you to spot problems in claim submissions or help prevent write offs in the future.

Monthly Tasks

Why perform a task only monthly? Normally this task is tied to the monthly financial statement or closing the books and has a "matching" element to it. For example the Advertising Budget is tied to the monthly advertising expense on the financial statement. Monthly tasks are also internal control issues – things that need a "check up" or reconciliation. They can also be planning items for budgeting and cash flow.

AP - 1099 track vendors

Keep track of vendors that require 1099 forms and make sure you have proper documentation. The easiest way to do this is to create a report of payments over the required tracking amount and to have a process to carefully screen when a vendor requires a 1099 form. Generally corporations do not require the form unless they are attorneys. Check each year the requirements, since they can change.

AP - Maintain authorized vendor list

Vendors are like AR customers and should be approved by management. If your DMS provides the field, indicate who approved the vendor. You should make sure your approval process includes getting the best pricing for the services. I also recommend approving vendors that buy from you. Run a report of the file and make sure the "authorized by" field is completed.

AP - Prepare Cash required estimate

At the end of the month, run a report of the AP schedule balance removing any debits (like the factory account) and other non-cash payments to show an estimate of what will be required by the 15th of the month. Using the AP balance can be inaccurate if there are large intercompany balances, or full note amounts. Try to keep this schedule only for the accounts that you must pay in full each month.

Advertising - Manager Budget

Advertising is a perfect example of a budgeted account. It must have a budget and purchase orders created. It is the 2nd most important item to perform an actual vs. budget after sales commissions. If your sales managers are having difficulty sticking to their budget, move the purchase order function to the controller and do not allow advertising without a purchase order. Advertising is not a variable expense. At the end of the month it does not go down like commissions if you sold fewer units. Consider it a fixed expense and budget a fixed amount. If you have to be the one to control the purchase orders, then do it. The attached file should be completed by the general manager, sales manager - or marketing department.

Advertising – Accrue

The advertising expense should be accrued at the beginning of the month based on the budget provided by the sales managers or general sales manager. A purchase order for each vendor should be created and it should be made certain that advertising invoices do not exceed the amount of the purchase order.

Advertising Budget Vs Actual Report

Be sure to match the expense period to the customer traffic period that the advertising represents. For example, if the advertising is for the last weekend of the month the dealership usually will not get the invoice until the following accounting period. Be sure to accrue for the expense in the same month that the advertising occurred.

AR - Review credit limits

Run exception report of any AR accounts that over or near credit limit. The parts department normally gives the potential charge customer a credit application which is completed and returned to the accounting office. You should have a policy in force with regards to required payment history, credit limits, and periodic reviews. Most DMS systems have a method to review sales and payment history. Careful attention should be paid to how many returns are made, gross profit percentage, and payment dates. If for example, they have established credit for a body shop that returns over 50% of their sales, has a gross profit percentage of less than 25% and is always 60 days late on payment, then this account should be reconsidered.

AP checks – prepare

I think AP checks should be prepared as close as possible on one day. Since they will all be signed and released at once, then try to print and post all at the same time so they don't mess up your cash in bank balance (by holding them.)

AP statements - reconcile

I don't believe in trying to get all the missing invoices when you are reconciling AP statements, rather pay the statement and indicate the ones that you need copies of for next month. Spend that extra time trying to figure out why you are missing copies – compare the ones that you get faxed to the originals and see if they are for sublet, warranty – and maybe the original got stapled to the repair order. Are the missing invoices signed by one particular person? Find out what they are doing with the originals and try to fix the process. The fastest reconciliation comes when the balance in the account matches the statement and you don't have to reconcile! Put your email address in the statement to get them emailed or faxed to you, for example; "missing invoice 12345, please email to sandi@SandraJerome.com or fax to 503-715-5600."

AP checks – sign

It is a good practice to have the two people that sign the AP checks to meet in a room and have a signing meeting. That way they can discuss the payments instead of one person signing and then having the 2nd person interrupt them later with questions. You can also do this during a manager's meeting and have the other managers attend to answer questions.

AR - Collection guidelines

Once a month, find out what are the legal collection guidelines in your area regarding how often you can contact the customer, when you can call and rules regarding contacting at their employment. Issue these guidelines in a memo and put in your employee handbook because almost anyone can end up calling a customer for missing funds.

AR - Send copies of invoices for any invoices not paid

As soon as you receive an AR payment check, try to notice any missing invoices on the payment that require copies. If you have been good about getting email addresses for the AP clerks, it should be easy to right click on invoices and email to the person that needs them. If you have a network fax, you can do the same thing.

AR Send out statements

Hopefully we'll be able to email AR statements, but for now, it seems like we must print and mail these statements. If the weight of the statement allows, try to include a flyer about your latest sales vehicles, rebates or service specials coming up. Don't be afraid to ask your managers to "hand deliver" a difficult AR account's statement.

Bonus - compare to pay plans

If you are using something like my Pay Plan Wizard, then you should be able to easily reconcile the amount of the bonus to their pay plan – since they should be on the same sheet, but otherwise take the bonus calculation spreadsheet and the original typed pay plan and compare the terms to make sure you are picking up the right amounts to pay your managers from – especially if there have been big financial statement changes.

Breakeven Report

A breakeven report tells you how many retail units that you need to "breakeven." It uses a simple concept; if the gross profit from fixed operations covers your fixed and semi-fixed expenses, then you actually make a net profit on the first vehicle you sell. This net profit is the gross minus commission expense. If your fixed gross profit does not cover your expenses, then you divide those "unabsorbed" expenses by your net sales gross profit per unit and that is how many you have to sell before you breakeven. When I was a controller, I'd do this calculation daily as a form of a "DOC" and let my GM or dealer know when we had "breakevened" and then he'd know the rest of the month was profit. He'd come in each day looking for that note on his chair – and hopefully it came before the end of the month!

Cash Budget Vs Actual Report

The first step is to get out your financial statement and go through the assets and liabilities and enter in the actual column. Next think about a budgeted amount of each account starting with cash. How much would you like it to be? In the example, it is really a large negative which might be making your job difficult. Enter in the budget column what would be more reasonable amount. Next, look at Contracts in Transit. You might want to use the Frozen Asset spreadsheet (in free downloads,) for this to determine what would be a reasonable amount. Continue down the financial statement making your wish list. I find that so much hides in AR and factory receivables. Often dealer, employee, and intercompany debt ends up here and that eats up your cash. Knowing about it can get these amounts into the correct account – or even better yet; square up your intercompany balances if that is running you out of cash. Obviously, if parts inventory is currently $631,241, then you can't budget it down to $25,000. If you've run the parts management report and he has $50,000 in frozen inventory that is not turning, it might be time to work on that to free up cash. The Internet has been a great resource for parts managers to get rid of that old inventory. You might spend your most time in Used Inventory less floorplan. I've worked with controllers that have said, "I floor everything in used, so no money there," but you might have a lot more equity than you think. Why? First, your floorplan provider might not let you floor everything like older vehicles and certain makes. Find these vehicles and figure out a solution. Run an analysis of the vehicles you can't floor and determine if they make enough profit to keep them – or maybe you should have a policy that these are wholesaled on day 1- instead of day 60 at a big loss . In new inventory, you might be building up equity by forgetting to floor dealer trades. Even though it was your trade and it should of sold – sometimes the deal doesn't got through and you end up with a new unfloored vehicle. Let's move onto liabilities. You might wonder, what do liabilities have to do with cash? That is easy, it is like your credit cards – the more you put on credit cards each month, the more cash you have. Not a good thing – but an easy example we can all relate to. A big source of interest free cash is Accounts Payable. Most vendors don't charge interest. I have been at dealerships where some vendors are coming by for weekly checks. That is terrible since you end up with 4 times the work – and poor cash flow. Find these vendors and get back to our industry standard of the 20th for payment. Yes, you should reconcile vendor statements starting on the 10th, but you don't have to release the checks until the 20th. A cash vs. budget doesn't do any type of footing or balancing (other than the actual column to make sure you got everything from the financial statement.) Since you're just making up the budget column, it doesn't need to total or balance – rather you're trying to figure out what is reasonable for these accounts and then compare them to the actual. We use another spreadsheet to do a more formal cash flow statement that determines exactly how your beginning cash became your ending cash – but that doesn't budget – it is more an "after the fact" analysis. I'd rather you determine how much you should have in each account in this exercise and wonder, "why is it different?" It is an exercise that we do when we start a new company and try to figure out how much we need to borrow or fund the new company – so it is also a good exercise even after many years of being in business.

Cash Sales/In and Out Audit

Close out in the middle of the day once a month and perform a cash in – cash inventory audit. Wait for a quiet time, like 3 P.M. and bring the cashier a new cash drawer with a normal day's opening balance. Take the existing cash drawer and paid repair orders and close out the day. Do the reconciliation yourself by stapling the checks and credit card slips to each repair order. The ones left over should be marked cash. Total the cash in the drawer and see if it covers the tickets that were taken in. If it doesn't bring the cashiers into your office (make sure you have coverage that day) and discuss what you have found. Show them the cash that you have and the stack of cash tickets. Ask what happened.

Cash Stretch Estimate

Once a month, determine how many months your current cash will last at your current "burn" rate. It is better to find out months in advance if you'll run out of cash if you keep losing money, increasing inventories, or receivables.

Chargebacks -Analyze

The number one reason for chargebacks in the new and used departments is an internal repair order that were missed when the deal was worked. Since I've never seen a customer purchase a vehicle that was being worked on in the shop, that means that the work was policy (done after the sale,) the due bill wasn't set up properly, or the repair order was still open when the car was sold. This means you probably have one of these three problems – or all three. Let's start with policy work in this issue. Policy work happens because there was a misunderstanding about the vehicle. Either the dealership misunderstood what condition it was in when it was taken in trade or purchased – or the customer misunderstood what was covered in the sale. Your first step is to quit hiding policy work. If a vehicle is repaired after the sale, it is policy. The key word is repaired. That means "fixed" and not merely equipment added as part of the sale. The hiding comes from having slush funds and packs. The whole reason why we create a monthly operating report is to let management know how we're doing and if we sweep these under the carpet, we'll never know how what really happened. I was at a dealership where the CFO suspected there was a lot of policy work, but didn't know how to find out how much. I ran a report
of his used vehicle sales for the past 6 months. Next I ran a report of all repair orders for the past 6 months and matched them by VIN. Using a Vlookup function, I matched the VIN# from the sales file in F&I to the VIN# in the repair order file. I then created a formula that found any repair orders done AFTER the sale date with an internal pay code. Make sure you sort the repair orders by date to find the earliest

date first in the list. If you do a lot of internals before the sale, you'll need to use a combination of Index and Match to find ALL the repair orders for a used vehicle. Using the "From" field do an analysis. How much average policy and reconditioning do you do per used vehicle and from what sources. Here's what we found; Trade in from New vehicles, average - $934.92 Trade in from Used vehicles, average $243.32 Purchased vehicles, average $1134.34 Used Vehicle Policy and Reconditioning Inline used, average $634.22 Other makes, average $1593.03. For some reason, the trade in vehicles from the new vehicle department, especially those of other makes had high policy and reconditioning costs. This dealership had a strict used vehicle policy that all vehicles had to go through the shop – the used vehicle manager could not send out vehicles without an open internal and sublet charge. It was great internal control and eliminated the need for purchase orders in that department, but at what cost? If they gave back the purchase orders to the used vehicle manager, he had a habit of doing a lot of "under the table" repair deals but they were giving an open checkbook to the service department. The other shocking discovery was the cost of reconditioning for the purchased vehicles. When this was analyzed, they found that the buyer was paid on a "per vehicle purchased" plan and he bought anything that he could – regardless of the condition. Further analysis showed that purchased vehicles had the lower gross profit average. This dealership fired the buyer and the new buyer was paid on gross profit on purchased vehicles. The next step was to have a policy to wholesale most "other makes" since they didn't have a good solution to stop the high reconditioning and policy costs. The last step involved a trade in from the new department. When those came in stock, the new department just tossed the keys to the service department, compared to a used trade that had an internal opened by the used vehicle manager that specified exactly what was to be done to get it ready for sale. They changed that process to enable the used vehicle manager to get the keys and authorize the reconditioning. For many of you – there is no way of knowing if you have these problems because we either hide policy work in slush funds and reconditioning ends up being a cost of sale instead of identified on each deal. Since the manufacturers are requesting that reconditioning be reported separately, you might start a process to account for this. You could put the purchase price of the vehicle in one account and all reconditioning in a sub account in the next column. Most DMS systems combine these two accounts when reporting costs – but if yours doesn't, don't do this since it is more important that the managers see the total cost than a breakdown. If you'd like to see your current inventory broken down by cost and reconditioning, then run a report on your schedule file and include the source for each item. Bring this into a pivot table and the control number goes in the row. The source goes into the columns, and the amount is your data. Use the account number as the page to filter. The new, used, and purchase journal sources are your costs. Make sure you use a vehicle purchase journal for adding inventory that is purchased and not the check or general purchase source. The other journals; parts, repair orders, checks, and general purchase orders are you reconditioning costs. Find the slush funds and either eliminate or analyze each month.

Commission Expense – Analyze

Each month, you should analyze commission expense with these steps; Step 1 – Determine your metric (goal) for sales commission expense. Step 2– How does your pay plan really pay? If the gross is $1000 and payable gross is $500, then a commission rate of 25% is really closer to 12% (divide actual commission by gross.) Step 3 – What is the expense from car deals? That means you'll need to break out your car deals. This is a tough step. First download your new and used journals or query your journal files into Excel and isolate the gross profit and commission for every deal. Next break them out by calculated commissions (those that were paid per the pay plan) and minis (those that were paid a flat amount due to a low gross profit or loss.) The problem came from the minis. One problem is now found. You must accrue sales commission on the car deal posting for this step to work. Step 4 – What are the other expenses charged to that account? Add up all the spiffs, bonuses, and minimum wages. Look at the various amounts and see if you can budget to come in line. All of these are controllable – make a budget for spiffs each month and only allow a salesperson to be paid minimum wage for a set amount of months before providing more training, coaching – or terminating. Step 5 – Apply the metric to your gross profit and actual commission and find out how much profit you're missing the target with by not meeting the metric. If your biggest problem is mini deals, you need to analyze each of those deals. There are tasks and tips for that!

Customer Complaints/Legal Issues – report

Maintain a Word document of all ongoing legal issues and complaints. As each complaint letter comes in, enter into a table in Word or Excel with these fields; Customer name, Assigned to, Outcome, Notes.

Customer Retention report

A customer retention reports tells you how many customers have been in for service during the past 2 years that you sold to. First, get a count of all vehicles that the make equaled what you sell new (I'm using Toyota as an example.) This count is done from the vehicles VIN# database. If a husband and wife each own a Toyota, then there would be two vehicles in the file – which is correct since that is the potential of two new sales – and there are actually two customers. The next count is of this database – how many did the dealer sell? I did this count by the make being Toyota and the stock# was not blank. Next we have two key retention items with the last sold and last service data. Calculate the rate based on the total counts. Compare these totals monthly to see if your rate is going up and down.

Customer Service Report from Factory

Even if you have a CSI or QCP Manager- make sure you get a copy of the monthly report from your factory of customer satisfaction results and put that in your month-end folder (or attached to the Done) of this task. Often you want to compare one year to another and the factory has limited storage of this data.

Credit Bureau Fees- Check Billing

Each month check this billing to make sure you are paying the lowest fees possible and your DMS setup that the minimum amount of reports are being run. I was once in a dealership that was running all 3 reports for EVERY customer although the sales manager told me they only needed to do that when the customer had very little credit or was from a certain state. I found the setup in the DMS system was to run all 3 automatically. I changed it to default to the one they liked the best and it saved them thousands per month.

Credit Life AH new policies - pay

There are two ways to get billed for Credit Life and AH; a transmittal – where you send in all the policies, listing the amounts and a statement where the company bills you based on policies sent in during the month. Before paying, it is important to check for any problem deals or unwinds because it is much easier to find these before you send in the funds and avoid a cancellation and have to wait for a refund. Some companies let you net out the cancellations.

CRM - backup CRM database

If you are using two products for customers; your DMS and CRM system – make sure there is a backup being created of the CRM database with the latest contact notes. If you can't back this up locally, or perform an export of your valuable data, then consider changing CRM providers. Check out DealerTeam/DealerStar.

CRM - make follow-up contacts with sold customers

Most manufacturers require that you made some contact with sold customers; a letter, phone call or email. Find out who is responsible for this and the method that is it being monitored.

DOC/MIS - modify for new account numbers/info

Once a month, make sure your, trial balance, income statement, GL, DOC and Financial statement match to the penny. If they don't, then there is a good chance that an account number was added that didn't get put on the DOC or Financial statement. Most DMS systems have a report to find these accounts, so try using that first.

Do Not Contact Guidelines

Check the latest guidelines for contacting customers – what is allowed to be sent via email, SMS, phone calls and make sure all employees know the rules and that all 3rd party vendors are following the guidelines.

Email Collection – Check Process

Spend a moment right now and visit your website. Are there at least 5 places where customers can give you their email address? Make it easy for customers to email you with questions. Make sure you have these items on your website. Want to know the status of your vehicle in our service department? Click here to email our service receptionist (or a simple form to complete.) Need to make a service appointment. Complete this form. (Make sure all forms have a required field for email address.) Want to know about the current recalls on your vehicle? Click here to be notified when a certain new model arrives. Used Vehicle Collectors? Click here to be notified when a certain year, make, model comes on the lot. Complete this form to receive a free appraisal estimate of the value of your used vehicle. Complete this form and I will find out how much you owe on your trade today.

Dead Deal Process and Report

The bottom drawer of the salesperson's desk is not the right place for dead deals. When a deal becomes dead, have the paperwork turned into the BDC and bring them back to life later. At least get them into your CRM program or a database before your DMS system deletes them. Check your process each month and run a report for the sales department. Why do deals become "dead deals?" The most common reasons are "financing" and "bought elsewhere". For those deals that couldn't be made due to financing – it's normally not a permanent situation. If they were "buried in their trade" – maybe after a year they have some equity. If their credit was bad, it might be better now or you might now have some special financing options. If they didn't have enough income, maybe a new job or raise has changed that. Even if they bought elsewhere – they might not be pleased with the purchase. If they are in paper format, find those old credit applications and write-ups – enter them into the database and make sure they are being destroy or stored properly according to the Safeguards Rule.

Denied Service Work

I'm sure you have some customers that have either cancelled their appointment – or declined service, so with the summer months coming up (when it seems that many have more free time) and a time when everyone wants their vehicles to be in top shape for summer driving – why not be like my dentist and go after those incompletes? It's easy to find your incomplete appointments and let's hope that you're using a field in your database for declined service, so a report is fast and easy. Run the report and email it to the advisors monthly.

Employees – Get Reviews from other Managers for their Staff

The payroll clerk should provide each manager with a list of all their staff coming up on their anniversary date next month. This gives the manager time to observe the employee and will enable them to write a better review during the month.

Employees – Perform Reviews for Office Staff

The payroll clerk should provide you with a list of all staff members coming up with an anniversary next month. Both reviews and bonuses are a way to not only let your employee know that you appreciate them, but since we are a profit generating industry – they have the added benefit of making more money. How can an annual review help you make money? It is easy – tell your employees what you need from them and it is human nature to please. I know this is hard to believe as you look out on your sales force or staff – but it is true. Nobody goes intentionally into work today with the objective to do a bad job. Employees do not like making mistakes and certainly do not enjoy you pointing them out. How do mistakes happen? The biggest reason is that employees didn't realize that it was important to careful. A review is the perfect time to go over these issues and put the greatest emphasis on those items that matter the most for that position. Let's take two examples; a back counter parts employee and a service cashier. Using a review form that rates their job functions on a scale of 1-10, here's how the session should go. "Sue, I'd like to show you the new method we will be using here at Sandy Chevrolet to perform your annual review. As you can see, you almost have perfect attendance and we consider a high expectation of your position – it rates a 9 on a scale of 10. Another high expectation is that the daily cash reconciliation equal zero. Unfortunately based on the input from the office manager, you score very low. She has provided me with this report of a random two week period. You only had one day where the cash reconciled to zero. Most of the days where it did not, you not only were considerably off, but it took you hours to come to that amount. I feel that this issue has directly contributed to your low score on overtime. What I would like for you to do is go back to your desk and list out some of the reasons why you are having difficulty reconciling cash sales and please list anything you think we can do to help you. Based on this review, I cannot recommend a raise at this point, but if you can bring this score up to a 6 in the next month, then I'll recommend a small raise and if you can reach the goal of a 7, then it would be a much larger raise. I'd also like to compliment you on your friendliness to customers, but you'd have a higher score there if you would extend that friendliness to the other employees especially the service advisors. For our office staff, the other employees are considered our customers. You have a neat work area and you scored very high, but we do have some issues with your attire and I'm going to circle the page in our company handbook where these issues are addressed." See how easy that is? Your objective is for that cashier to do a better job of controlling the cash drawer and reduce her overtime. Improvements in both these areas mean more profit, so this review can make you money. Let's look at the parts counterperson's review. Again you would compliment him on his attendance and conformity to company rules but point out the low fill rate and cooperation with technicians. If this back counterperson were to hit the goal score by improving your fill rate and increasing technician productivity – that would mean more profit. He also needs to get his overtime under control and that reduces expenses. I recommend using a review form that rates the employee. You can make it more specific if you want by providing a raise percentage for each goal step. Doing a review is uncomfortable for all participants, so if you stay focused on your objectives and their goals instead of turning the session into a painful discussion. Write down bullet points to help you through the review and provide these bullet points along with the scoring page to the employee. I'd leave room for them to enter their comments.

Employees - Change rates per review

The payroll clerk should get the completed reviews and then change the employee's rate per the review in the computer before the first payroll after the effective date.

Employees - Employee databases – check

Download the employee database and check to see if these fields are accurate; active status, last review, rate change, address, phone, date of birth, termination date. If employees are getting added without complete data – or if terminated employees still show as active, improve your database maintenance.

Employees- Update employee handbook as necessary

Keep a folder on your computer of all memos and determine which of these should be updated to your employee handbook. Go to your state association and NADA website and look for any new regulations. Try using a virtual employee handbook that is on the desktop of each computer for the latest one, and print new ones on demand (instead of printing each year.) If you do keep one on the desktop of each computer, make sure your signed agreement refers to that file.

Employee insurance - pay statement

I've seen two ways of paying this statement. The first one has the amount that the employees pay going as a credit to an expense account and then the full statement debits this. The other and more accurate method is to debit the employee AR for the employee portion of the premium. Obviously the first method is much easier but I imagine that it is very easy to miss charging an employee for their insurance, especially Cobra former employees.

e-Newletter Create and Send

Many of your customers are spending more time in front of their computer than in front of the television, listening to the radio, or reading the newspaper. An e-newsletter is a great way to reach them. Yes, many of them will be clicked on and deleted (just like my old Excel Tips and then my Task Tips) but I actually get over 30% of mine opened. I've used both Constant Contact and I-Contact for my Excel tips. The benefit of using these services is that it is polite. The recipient has the opportunity to opt out without any effort on my part. They maintain my list for me. I use two since some ISPs block one or the other. You can have various types of newsletters – something rather specific like my Excel Tips or more general about vehicles. The key is that it contains something valuable for your customer; a coupon or special sales price. It also needs to have the purpose of getting your customer into your dealership. Special events like owner's clinics or service specials attract customers. Before you hit "send' ask yourself those two questions – why would the customer read this and does it give the customer a reason to come see us? Now, I'm starting a list of people who want to know when my latest books come out.

Extended Warranty new policies - pay

There are two ways to get billed for Extended Warranties; a factory debit to your parts account or where the company bills you based on policies sent in during the month. Before paying, it is important to check for any problem deals or unwinds because it is much easier to find these before you send in the funds and avoid a cancellation and have to wait for a refund. Some companies let you net out the cancellations but others do not allow a "flat cancel" after a certain amount of days.

Expense Account Analysis

Depending on your computer system, detailing your expense account usually involves an inquiry of the account. If possible get all the accounts at once and then cut and paste the detail into Excel. For example if you have R+R, you'd use job 0504 (use next to move to the next account), 0367, or 0368. For ADP, you can use AGRI. If you have some of the newer programs like ERA XT reporting or ADP's w.e.b.Suite series – or one of the new PC-based DMS systems– detailing accounts is much easier. You can normally cut and paste from the screen into Excel. There are also add-on products like ADP's AGRE with PBX to get a report of base accounts. In addition, you can write a query or RPG from your detail files; GL.JE.DTL for ADP with the account type of "E" or the file JOURNAL for R+R. For other systems use their reporting tool from the GL detail file and either use a range of accounts to get all the accounts in your base account, or pull all with a type of expense into Excel and you can filter in that product.

Expense – Item Analysis

Using the procedure in my Super Controller 2 Guide – Expenses and Employees, pick just one item in one expense account an analyze it.

Factory Monthly Parts Statement

This might be a little too difficult for a beginner accounts payable clerk, especially someone who came from another industry and wasn't used to all the various types of entries on this statement. If you have an office manager, this is an ideal chore for that position to learn more about the whole dealership - and remember, it is not part of the close. There should be nothing you need on this statement to close the books – everything should already be posted from daily invoices.

Filing – Car Deals

I'm a firm believer that car deals don't need to be in the office. Most of the information that we need is on the DMS, but if you must file them – then monthly is a good plan. The current month's deals should be kept in a 4 drawer file cabinet next to the billers' desk until they are complete. A complete deal means the title work is done, the funds have been paid by the customer, all rebates collected and reserve reconciled. Once this has happened, then it can move from a pending drawer to "ready to file". They should always be kept in Alpha order to make it easier to file. Just have both drawers have A-Z tabs and put them back where they belong. When the file clerk comes in to file the complete deals, they can be pulled out in order. If you're doing the "flag" system of car deals, then the file cabinet can be in the middle of the room.

Fixed Asset Purchases Report

Each month prepare a report of all fixed assets that were purchased unless your DMS has a fixed asset module where you can add them monthly, then run a report and email that to the dealer/GM or corporate office. Be sure and include who authorized the purchase. For large purchases, adjust your standard entries to account for the increased depreciation.

Four Hours Free Time Off

Determine for the next month close the deadlines that need to be met for the office to earn their 4 hours free time off for getting the close done according to the schedule. Post it on your door before the last day of the month.

Floorplan Reconcile to Lender's report

I've found that downloading your floorplan and using Vlookup saves about 4 hours of reconciling a medium-sized floorplan statement. I was delighted to discover that you can now download your GMAC floorplan outstanding - and I'm sure most of the others also allow it. GMAC has added a nice Download button to the Search Vehicle Inventory screen. If you cannot find a way to download your outstanding floorplan, you can often search or browse your current outstanding charges and cut and paste them into Excel. Next use the Data, Text to Columns to parse the data into individual cells on your spreadsheet. I find that using mini schedule download is the easiest way to get your "book" floorplan outstanding into a spreadsheet and again, you might need to parse that report. Use you Vlookup function to look up the VIN# on your outstanding floorplan report to put the amount floored per your vendor (like GMAC) next the balance on your floorplan statement. Use the same function to look up the VIN# on your mini schedule to put the amount on your books next to the outstanding amount on the vendor statement. When Excel cannot find that VIN#, it will enter a #N/A, so you'll need to enter this formula to find any outstanding amounts on the two reports; =IF(ISNA(H2)=TRUE,G2,G2-H2) - this means that if the amount in H9 is a #NA, then put the full amount in the previous column, G2, otherwise subtract the difference. Let's say that column G is your schedule amount. If they Vlookup function does not find the VIN# on the GMAC download, then it gets a "#N/A" so the whole amount in G2 is outstanding. Let's say it finds it, but it's floored for a thousand less. Then the difference between the schedule balance and GMAC balance is outstanding.

Forms - maintain sample book and vendor source

I like to keep one master book with an original of all form that the dealership uses and where we order the form from and the last price. This enables new vendors to try and beat our price by enough to make it worth our while to change vendors. I don't change for a small amount, especially if the vendor purchases their vehicles from us or has service work done – or buys parts.

Forms - take inventory of supply/order

Once a month, I give the book to a clerk to go through the dealership and get a count for each form so that we don't run out of an important form. Any that are getting low are ordered.

Frozen Asset - AR

This is another area where you can have cash "tied up." Most service customers should pay by cash, check, or charge cards. If you have gotten in the habit of setting up a charge account for good service customers, it is time to start having the customer use their charge cards instead. Not only will the customer earn frequent flyer or other bonus points, but the dealership will avoid collection problems and possible damage to customer retention. The other area is the wholesale parts business. Be extremely careful when building a large wholesale parts operation. Not only do you tie up inventory in the parts (often sold at a very small gross profit) but you grow a huge AR balance. Each month perform this analysis to make sure it isn't getting too large.

Frozen Analysis - Contracts in Transit

Each month you should not only look for past due items in contract in transit, but also do an analysis that might provide a new insight into contracts in transit. Often we are not concerned with the schedule since no items are over 30 or 60 days old, but in reality we have too much money tied up in Contract in Transit. By looking at the balance as a whole and then comparing to how much you should have – can spot problems in the submission process that could be improved by maybe having the F&I managers pull the contracts and submit or switching to a more paperless – eContract system.

Frozen Asset - Warranty Claims -

Similar to the Contract in Transit, the warranty claim schedule might not have anything over 30 or 60 days old, but is too high of a balance. If the warranty repair orders are part of the customer or internal repair orders then delays in submissions could be caused by a delay in closing the other portion of the repairs. Other delays are caused by improper submission to the factory or waiting for an approval from a district manager. The schedule may also contain warranty payment shortages or denied claims. By performing this analysis monthly, you can first determine if the balance is too high and look for ways to streamline the process so you can get paid quicker.

Frozen Asset - Finance Reserve

With many lenders now paying reserves when they fund the contract, there never be more than a 30 days supply of Finance Reserves. There are a few reasons why the supply would be more than 30 days worth; 1. The finance company pays only quarterly. 2. There are adjustments (shortages) in the schedule that have never been adjusted. 3. The chargebacks have not been accrued. 4. A big push of deals at month end. By performing this analysis each month you can find out if there is some changes you can make. I once called a lender that was holding a huge reserve and got most of that released since we no longer did many contract with them.

Finance Reserve - Schedule or Account Clean

Hopefully you have this account scheduled rather than printing a detail monthly transaction list for the finance reserve and then manually writing the amount credited by the finance company next to each item. This is one schedule (like warranty claims) where you might want to use automatic write-offs if your DMS has that option. If the difference amount is less than $2.00 or -2.00, then an automatic write-off can be generated. If not, I find that if you download the summary schedule to Excel and then sort by amount, it is easy to write your journal voucher by reversing the sign in the next column.

Finance Reserves – Chargebacks Analyze

Each month you should do a random check of the chargebacks for repos and early payoffs to make sure you are being charged back in agreement with you contract with the bank. I have found that some lenders will charge you on a repo for any collection charges, late fees, insurance not paid, etc. Make sure this is not happening - unless that is part of your agreement. Have your F&I manager calculate the payoff and then check what the lender has come up with.

Finance Reserves – Held Funds Agreement

Each month, check the balances that your lenders are holding in "reserve." Often the loans have long been paid off and you no longer do business with them - see if you can get that back. If you did sign up for the "no chargeback option" from a finance company years ago, then the percent of the outstanding balances probably no longer applies and you can get the money back.

Gas for demo rides - verify amount of gas policy

Go through the gasoline bill for the last month and find all the slips for demo rides. Check the amount of gas that was put in the tank – normally 5 or 10 gallons and look for any "fills" for the same stock number – and units not sold and used vehicles.

Go Green List – Update

Since going green is so popular, keep track of the ways your dealership is green. Find out from your shredding company how they handle that material and put that on your list (most recycle the material or make it into packing products.) Add the little tag to your email to remind people not to print emails. What about your waste oil? Do you have a recycler pick up? How about tires and batteries? Find out what a good job you're probably doing to recycle those products. How many of your new vehicles get great gas mileage? List all of those items on your website or in a special email newsletter.

Identity Theft Prevention Program – Audit

First you should approve the initial written ITPP (responsibility of the board of directors, dealer, CEO, or other member of senior management) 1. Assign responsibility for development and administration of ITPP (typically controller or office manager) 2. Train employees on policies and procedures developed in the ITPP and the importance of compliance 3. Oversee service provider arrangements to ensure that activities related to "covered accounts" are conducted in a manner consistent with policies.

Income Tax – Estimates Due

Make any tax estimate payments for income tax.

Insurance Claims - Listing of all Losses

Create a report that lists all losses for the month. When you are entering into your DMS, you'd use the category or Description, "Insurance Claim" to filter all tasks to make this report easier. The report should contain who was involved in the loss, filing information and who is responsible for filling the claim.

Internet Leads – Check billing for duplicates

Download your billings from Internet (most providers have a copy of your bill online) and make you are not being billed twice for the same customer by using Excel or a database to find duplicate billings.

IT Cost Analysis

If you've looked at your DMS billing lately it might be shocking how expensive your DMS system is getting unless you're one of the lucky ones that switched to an alternative low-cost "one line" DMS provider. For others, this billing might be bad news, but it gets worse – your true IT cost is normally double of what your billing is from your IT provider if you're using a Tier one provider and four times of what you're paying an alternative provider. Let's say that you're paying ADP or R+R about $9,000 a month. There is a good chance that your true IT costs are $18,000 a month! If you're paying AutoSoft $1200 a month (for ASI and F&I) then your true IT costs are closer to $4800 a month. How can this be? Let's discuss how to make a truly "apples to apples" for everyone. The first cost on the analysis is something we really don't expense to our financial statement under IT costs. This is the expense of buying everyone new PCs, upgrading our DMS system, or even buying new software. Most of us "dump" that cost into fixed assets and deal with it at the end of the year. Sometimes we might make an effort to start depreciating the purchases during the year, but most times we leave it for the accountants to handle and they will normally expense it as a 13-month adjustment. Does that mean that all this stuff that you're buying doesn't really cost anything? Been wondering where all your cash went? This is the little hidden cost that really adds up. Next is– PCs. This picture is only a tiny portion of the PC section since we also have file and print server support, email, Microsoft software, etc. This is where your biggest increase over the past ten-twenty years has happened. About 20 years ago, 50% of the users still had those ugly green/blue screens; dumb terminals. Now it's hard to find one left in a dealership. Everyone has a PC and that means Internet access, virus software, and Office. PCs are hard to support and even harder to control. Do all your employees really need Internet access? With sites like Facebook and Twitter booming, the world's workforce may find it increasingly hard to avoid such alleged time wasters while at the office. Often, leisurely browsing can simply be categorized as procrastination, according to researcher Jennifer Lavoie, who conducted a study earlier this year on online procrastination. Psychology Today blogger Timothy A. Pychyl wrote that in that study, "50.7% of the respondents reported frequent Internet procrastination, and respondents spent 47% of on-line time procrastinating." Next is the dreaded Printers. This is one of the few technology items that have moving parts and I can't tell you how much time I waste trying to get paper out that is jammed in my scanner or printer output. The best solution is bigger and better printers – and less of them. These "personal printers" next to everyone's desk are not built for the volume we put on them and most cost analysis will find that an industrial printer will use less toner and paper. Making employees get up and go to the shared office printer will reduce casual and personal printing. Servers are another thing that has been tossed into our lap by the new evolution of DMS systems. The server used to either be owned or owned and supported by the DMS provider. With many alternative systems and new versions of bigger DMS systems; the server is now yours to upgrade and maintain. Many times, projects come up that are beyond the capabilities of the average dealership IT department, thus the need for the next section of consulting. If you have an IT department, then you have to allocate their monthly costs (be sure and include taxes, insurance and benefits) between where they are spending their time (thus the need for the Vendor/Hours column.) In this example, the IT person costs $6275 a month and spends 20 hours a month doing backups of data. This works out to being about 12% of the cost, or $721.26. The last section in the spreadsheet is the one that we've been picking apart for years; DMS system costs, credit bureau fees, CRM systems, etc.

IT - Data protection - Safeguard/IdentityTheft/Red Flag

Perform a review of Safeguard/IdentityTheft issues with your customer database and other company data. Remember that your customer database might be in many places; the DMS, CRM, and other 3rd party programs that extract it. Make sure privacy notices are up to date.

IT - DMS - arrange for staff training

Training for your DMS needs to be an ongoing process – not just during the install, especially if you're going to become a task-based accounting office. Find out what training is available – usually online and schedule the classes for a small group of your staff.

IT- DMS Billing Analysis

Each month your accounts payable clerk should reconcile the DMS billing to the contract. To make it easier, you need to start with your original contract and enter those amounts into Excel. Next, take the first billing that you received on the contract and reconcile to the agreement. Now you have your base for each item. Sometimes you got a lump discount when you purchase the system and you need to find out how they applied the discount. Then read your contract and determine the amount your contract allows for annual increases. I always recommend no increase for two years and then a max of 2% a year. Of course now that I'm getting into the DMS business, I think 10% is fair! Just kidding. Next, take the billing a year later and enter the amounts and calculate the increase and compare to your agreement. Normally the billing department number listing at the top of the invoice will straighten out these "mistakes" and fix your invoice and all the ones in the past. No need to

bother your sales representative since he or she would need to call billing anyway and I find the
person in accounting is always easier to deal with! What about billing errors? Here are the most common; 1. Annual increase is higher than agreed to. Look at your contract and see how much they can increase your monthly support. I have found individual items that increased as much as 30% instead of the 2% that was agreed to in the contract. I've also seen increases mid-term, even two price increases in one year. 2. Hardware you no longer own. When you have an upgrade done or replace equipment, the old equipment doesn't come off your billing unless you ship it back, send a letter, and call the billing department. Take an inventory and make sure everything you're being billed for is still around. I recently found 6 modems on the billing but only one was connected. The other 5 were in a pile in the corner – placed there by the technician after they changed how some of the buildings were connecting. 3. Software you don't use. You'll need to go over these items during a manager's meeting or during a "walk around." I asked the service advisors how they liked their pricing guides and they replied, "what pricing guides?" They didn't even know they had the software and then when I showed it to them, they didn't like it and refused to use them. We sent a letter to get those charges and the expensive updates removed from the bill. 4. Time and Materials. Why? It's very unusual that you would have a time and materials charge unless it was your fault that the item broke. You'd almost have to take a hammer to the CPU for it to be your fault. Question these charges, it's often a technician's error on the ticket – or an "after hours" support call that was during your regular business hours and should be included in your monthly support. Read your contract. 5. Special paper, supplies and forms. Watch these items and before you just pick up the phone and reorder – ask if you need it? Do you really need pretty yellow hard copies with the lines printed on the back? Wouldn't plain paper work just fine? Even better, why do you need the hard copy anyway – if you have ERO or ESI? Couldn't the technicians type their notes on the computer?

IT- DMS - change forms as required

If your DMS company enables you to change your forms, it is not as hard as you think. The best place to start is to copy the old form to a new form name. Next look for fields that you recognize on the form and notice how the code determines where the field prints. Using this as your guide, just add the new fields or move the ones that you need.

IT- DMS - database duplicate customers

It is important to join duplicate customers into one record in your database to prevent duplicate mailings, phone or email contacts. The most common method is to use a search that finds the same last name and address. Of course small things like one record using St. and the other using Street – can prevent the search from working. The next level is a last name and phone match, then a first and last name and zip code match. Each time you're using other fields to try and determine if these records are truly duplicates. Next you need to determine which customer# is the one you "keep" and which one is the one that is joined to the other? I normally use these fields in this order; last service date, last sold date, and selecting the salesperson who is still working at the dealership.

IT - DMS - setup of F&I Interface

With a tight interface between F&I and accounting, deals can almost "autopost" themselves. If you make sure each F&I field goes to the right place in accounting and uses the right control number, you can save so much time for the billers. I've seen interfaces that were so bad that the biller deleted all entries and started from scratch. What a waste of technology! Sit with your billers and notice what they have to enter manually and change – then set up the interface to make it right.

IT- DMS - setup of Parts/Service Interface

The best way to find out if you have problems in the setup of your parts and service interface is to look at last month's journal vouchers (a little consultant's trick!) If you are always making entries to fix the PDI credits, warranty and sublet, and costing – then there is a good change you need to look at the parts and service interface. I make a copy of a parts ticket and service invoice and then manually decided where each amount should go; what account and control#. Next I check the interface and find the problems and correct them.

IT- DMS system - updates

Most new DMS systems "auto update" so this might be a task in the future that will get deleted. If you have one that still sends you tapes and disks to use to update your system, then make a plan to do this once a month, unless you start getting more than one. Often one update must complete before another, so you might have to break it down into different nights.

IT – Security of server, network

Perform a review of your network security and servers. Is the door to the server room locked? It is easy for someone to come into a dealership and just take the whole server! Walk around the dealership. Are PCs left unattended and logged in? Consider password screensavers and teach everyone the Windows Key and L – to lock the screen when they get up to get coffee.

IT – Backup of PCs

When was the last time you did a backup of your own PC – or important ones in the office? This is the question you get asked AFTER you have a hardware problem. The best backup is no backup at all! If you try to store most of your data on the company server and then if that is backed up nightly (I hope) you don't need to back up your own. I live in a hurricane area so I do 4 backups; one locally, one on the server, another on a fireproof server, and a third in a remote server. Fortunately, backing up your essential files is neither difficult nor time-consuming. In fact, the whole process can take as little as 10 minutes a week, and you can let Windows do most of the work. Click Start, point to All Programs, point to Accessories, point to System Tools, and then click Backup to start the wizard. Click Next to skip past the opening page, choose Back up files and settings from the second page, and then click Next. You might be tempted to click All information on this computer so that you can back up every bit of data on your computer. Think twice before choosing this option, however. If you've installed a slew of software, your backup could add up to many gigabytes. For most people, the My documents and settings option is a better choice. This selection preserves your data files (including email messages and address books) and the personal settings stored in the Windows Registry. I picked "Let me Choose" and selected my six folders of most important data and my Outlook PST file. On the Backup Type, Destination, and Name page, Windows asks you to specify a backup location. If you're one of those exceedingly rare individuals with access to a backup tape, the Backup utility gives you a choice of options in the Select a backup type box. No tape drive? No problem. Backup assumes you're going to save everything in a single file; you just have to choose a location for that file and give it a name. By default, Backup proposes saving everything to your CD or USB drive. Other choices are; 1.) Your computer's hard disk. The ideal backup location is a separate partition from the one you're backing up. If your hard disk is partitioned into drive C and drive D and your data is on drive C, you can safely back up to drive D. 2.) A shared network drive. I picked the 2nd drive on my network server. After you've chosen a backup location, enter a descriptive name for the file, click Next to display the wizard's final page, and then click Finish to begin backing up immediately. If you're disciplined enough, you can repeat the above steps once each week and perform regular backups when you're ready. If you'd rather not count on remembering to perform this crucial task, however, set up an automatic backup schedule for Windows. When you get to the final page of the Backup Wizard — don't click Finish. Instead, click the Advanced button, and click Next to open the When to Back Up page. Choose Later, and then click Set Schedule to open the Schedule Job dialog box and pick a time to perform your backup (I picked every Friday at lunch.) Easy! I'm sure this will change with each version of Windows, so just search on Windows Backup for the latest instructions.

IT - Software and System Updates

What version of software are all your PCs running on? Is it time to upgrade? What about browser versions? Some manufacturers, CRM, and DMS companies only allow certain browsers, so keep a list of these before considering updates. What about other software? The first step that you need to do is determine compatibility. Go to Start Programs and write down each program. Check the company's website to make sure your version is compatible with the one you want to update to. If you trust Microsoft, then you can use Windows Upgrade Advisor (search the Microsoft site.) Download it and run the Windows Upgrade Advisor to see if your PC is ready for the next version of Windows. It scans your hardware, devices, and installed programs for known compatibility issues, gives you guidance on how to resolve potential issues found, and recommends what to do before you upgrade. If performance is an issue, I don't recommend upgrading an existing PC – instead if you're buying a new PC and plan on loading the software you already own on it, then you can make sure by testing on your old PC if things are going to work.

IT - Technology Budget

Each month have IT prepare a technology budget for the next month. As technology costs increase, it has become more important than ever to create a comprehensive budget that includes your DMS charges, supplies, new equipment, repair, software, and personnel costs. Carefully watch overtime and see if you can overlap the shifts of technology personnel to minimize this.

Mini deals -Analyze

I don't believe in mini deals. If you have a vehicle where you can't make a decent profit, then it should be wholesaled. To find out what is causing minis, you first have to find where each of your vehicles with a low profit came from and why their cost was too high. For new vehicles – what happened? Who ordered them? Did they come from dealer trades? Did you add too much equipment or keep them as a demo for too long? For used vehicle – should these have wholesaled instead of paying that sales commission – especially if they weren't vehicle in your product line that could convert to service customers? Who appraised the vehicle – or worse – who purchased it? Doing this analysis each month, will help managers learn what is causing their low grosses and hopefully correct it in the future.

Mailers, flyers - Send out

Sending information by mail might be an old technology, but still very effective when it comes to certain campaigns. The same person that deletes almost all their email Inbox, might spend a few minutes walking back from the mailbox looking at your flyer about new models or a service reminder. If you've brought this task in-house to save money, try to make this more of a monthly chore and to a much smaller audience. Why send out 22,000 service flyers when 2,000 might be more effective if sent to the right person?

Manufacturer Guide - review and update

This guide is the policy and procedures that govern the relationship between your dealership and your manufacturers. Makes sure you know where the original copy of the guide is kept and who is adding updates – or is the full guide now available online. Don't wait for a dispute with the factory to get this guide updated – or worse search for it at an important time.

Marketing – Campaigns

Each month, there should be a marketing campaign for each department – unless they already have enough customers. At a manager's meeting, ask each department what will be their campaign for next month. This is important to record and later create a report of campaigns by month and the results.

New vehicle inventory vs. Sold report

This report compares what you have in stock to what you have sold. As an example, if you have 25 F150 pickups in stock and sold 27 last month, then you are short 2 units to have a 30 day supply. If you sold only 12, then you have one extra unit more than a 60 day supply. It also has gross profit on the report so that you can try to carry what sells in the least amount of time with the highest gross profit.

New Inventory - Days Supply

Days supply is much different than days old. Yes, you might have no vehicles over 60 days old, but you might have a 102 day supply based on your current sales. We sometimes get caught up in a thought of we need 7 million in inventory to have a good month. That is if you're selling 3-4 million a month. If not, you might only need 4 million in inventory. Since our new crop of lenders will be looking at this more like bankers instead of captive factory people, it's time to adopt a policy of watching your days supply.

New Inventory - Review ordering procedure

Who orders your new vehicles? I find that in seminars, most controllers don't know the exact person in charge of spending more money than anyone else. They are often shocked to find that the manager who is supposed to order vehicles has turned this over to a clerk, fleet manager, or even worse – let's the factory decide! There should be a written policy used by the dealership to order new-vehicle inventory. Either the dealer or general manager should be responsible for making the vehicle ordering decisions and these decisions should be based upon an inventory worksheet provided by the sales manager. The accounting office can help produce the information that is needed to make the decision. Use a worksheet as a guide to ordering vehicles. It is important to know what has sold in the past by model, trim, equipments, color – with the least amount of days in stock and the highest gross profit. The factory only knows what you sold – they don't know how long it took and at what cost to your gross profit averages. Don't let them decide what you should carry.

OFAC Specially Designated Nationals (SDN) – Review Guidelines

Make sure you are aware of the latest guidelines for verifying that you check the SDN list before doing business with a customer and get the information out to each employee.

Office supplies – order

Most large office suppliers and warehouses (Costco, Office Max, Staples) now have online ordering with free delivery. If you order from one place, you can keep track of your orders – but if you give this chore to someone who likes to get a good deal – that person can watch for specials and order when things are on sale.

Office supplies - put away/organize

The best place to store office supplies (for the office staff) is in their desk. After an order arrives go through the office with a basket and ask, "do you need pens, post-its, paperclips?" Whatever is left can go into the supply drawer. I don't think you need to spend so much time keeping it locked (and people asking for keys) if you keep it in plain sight in the middle of the office.

Order sales and service supplies (key tags, etc)

If you use the forms book, then you can also keep copies of supplies in there too like key tags, sale tags, etc. Once a month, go through the form book and take inventory and order whatever is low. Make sure you verify these orders with the sales managers – they might have stopped using a certain supply.

Parts Management Report

The parts department prints out a parts management report every month. It generally contains a line that clearly indicates the stock order performance. This figure is generated as a direct result of how the parts manager coded all the parts receipts (purchases received by the parts department) for the month. This report also shows their "memo" amount of inventory value and that should be compared the accounting balance each month. It will be off (look at the annual parts reconciliation to see the timing difference) but it should be off about the same amount each month. Looks at the aging of the parts - and see if there is an excessive amount over 12 months old. Notice the manual adjustments to the inventory. The plus adjustments should equal the minus adjustments. Excess minus adjustments could indicate a theft.

Parts inventory to Accounting

In addition to an annual inventory, it is best to compare the monthly inventory per parts (parts management report) to your accounting balances. Using the spreadsheet, keep a tab for each month, so if shortages to come up during the physical - you can possibly isolated when it happened.

Payroll – Monthly Tax Reports/Quarterly Reports

Complete any monthly or quarterly payroll or workman's comp reports as required. Carefully set up next due dates in Outlook or Google Calendar – whatever you use to keep track of important dates.

Physical Inventory – New and Used

Take your new and used physical inventory. Scan or write down each VIN#

Physical Inventory – New and Used - Reconcile

Reconcile Physical inventory to the new and used schedule (see website reconcile too.)

Payroll - garnishments, pay

Most garnishments are required to be submitted monthly and specify the due date. Check if you can now do an online submission (child support.)

Payroll - order payroll checks

Check your supply of these monthly. Try to move towards direct deposit and print on blank paper to save money. If your DMS system doesn't do direct deposit, why not look into a Payroll Debit card? Rather than issue a paper check, the employee's net pay is electronically transferred to an individual bank account instantly created for each employee. The account is then accessed through the
debit MasterCard issued to the employee. Some of the benefits for the dealership include; Significant savings on payroll expenses; No start-up costs. The cards are typically free to employer and employee; Eliminate hassle and expense of lost and stolen payroll checks; Eliminate paper payroll stubs - use electronic and/or phone pay stub system; Proven to dramatically improve employee retention; Eliminate off-site trips by employees to cash checks; Eliminate off-site and employees who are off work from having to come in to receive pay checks; Pay terminated employees immediately; Incorrect pay amounts can be corrected immediately; For the employee, some of the benefits (depending on your bank's features) include; Individual FDIC insured bank account accessed by their debit MasterCard® card; The bank account may be a qualifier for some grantors of credit; No charge to make purchases when using the card as a signature card or purchasing online; No monthly or annual fees; Free, live customer service 24/7 via toll free number; Cash from any ATM machine; Rent cars, make airline and hotel reservations, track spending online - 100% of all past transaction records are always available; Share money with family nationally and internationally at a very low cost; Get cash back at merchants with purchase (requires PIN transactions); Pay bills by phone or Internet; Reload prepaid cell minutes; Make long distance phone calls; Eliminate costly check cashing fees. Ask your bank today about payroll debit cards – your employee will love it. How to interface? Most banks want the payroll data sent to them in a csv file

– just write a download from your DMS system into Excel and with a macro create the export in the format they want.

Payroll 401k plan – delete and enroll employees

While doing the stat form for the financial statement or the employee add/delete report, it is a good time to delete employees from your 401K plan and add any that now qualify.

Payroll 401K plan - submit payments

If you want a shortcut in submitting payments, credit all payments to an AP account and let the 401K check get printed with the other AP check in the monthly run. Attach the check to the report that you print from the DMS system and you're done!

Phone Test

If you don't have weekend and night receptionists anymore – then you might consider adding back this position first when things get better. If you want to see if you need this position, just call your dealership and ask for one of your salespeople or say you want info about a car you saw in the newspaper this morning. If the call isn't being handled well (rings and rings – or the eternal hold) then this is the first place to your staffing expense dollars. There is a good chance you once laid off this position because she was always sitting there reading a book with nothing to do. A better thing to do is to make this person very productive without hurting the phones.

Postage - fill meter with value

If you're using an older machine, it might be time to upgrade to one that has new features like filling online and automatic upgrades of postage changes. Most are rented, so make sure you're getting the latest model for your monthly rental. Tie the account to a credit card and you'll never need to worry about this chore again.

Prepaid Accounts Reconcile

Prepaid balances are easy to creep up. If your insurance company asks for the full payment up front – why not ask them if you can pay monthly? Make sure there is a true savings for forms purchased 12 months in advance – 3 – 6 months might be better.

Property Tax –Reports and Pay

Check with your state/area for any required property tax reports and payments due. Carefully set up next due dates in Outlook.

Rebates - maintain master file of all programs

This is one area where I like to use a more manual method of record keeping. When a rebate program arrives, print a copy and put into a master rebate binder. The reason for this is that many manufacturers remove rebate program guidelines from the website when the program expires – so if you're working on an older rebate, you might not have the rules anymore. In addition, if you have a sales tax audit years later, then having the rules could help you avoid sales tax (dealer cash rebates are non-taxable – IF you can prove it was dealer cash.)

Rebates, End of Program Audit

At the end of each program end, run a report of all the vehicles sold that qualified based on year, make, model, etc. and then check it again the full report (normally obtained online) of all rebates paid during the program. This is especially important with Dealer Cash programs that might not be set up on a schedule.

Registrations – Temporary, Order and Audit

Check supply of temporary registrations and order. Perform a quick audit of book copy – looking for duplicates, voides.

Rent – Pay

Pay any monthly rent required.

Repair PCs, printers

If you don't have your own IT department, then the repair of PCs and printers should normally be a "swap out." The cost of a new PC or printer is usually much less than the cost to call someone in to repair. The hard part is determining if the PC or printer is truly broken – or if someone is not using it properly. In one dealership, the person who did this work was a technician and he merely flagged his work. Ask around the dealership and you might find a salesperson (check with the Internet manager) or parts employee that could be your "on call" repair person. Even if you do have an IT person or department, having an extra person for when printers get locked up or PCs won't start up – is great.

Resale Certificates

Check when new resale certificates need to be obtain per your state's guidelines.

Sales commissions - prepare for payrolls

This should be a much more automated process than it is in most dealerships. A complicated commission voucher system of paying commissions hurts your profits. The voucher forms cost money, the commission summary sheets cost more money and both are labor intensive and are full of mistakes (we always catch the underpayments, but the overpayments go by undisclosed.). If you pay commissions at 25%, then why not run a report of the car deals for last month, provide the totals by salesperson and then pay them 25% of the total? Sound too simple to be true? Yes, I'm sure there are a few reasons why this won't work, so let's address those.

Chargebacks; find out why you have chargebacks and stop them from happening. If repair orders are left open too long, then close them quicker (why are you waiting for the hard copy?) If charges are not getting added to vehicles – use an automated PO system and due bills. Chargebacks cost you more than just the overpayment of commissions – they have a negative effect on the sales force and that can hurt sales. There is no excuse for chargebacks – your sales managers need to know the cost before they sell the vehicle. This is a much larger problem than trying to make sure we recover the commission expense. Mini deals - I don't believe in mini deals. If you are forced to sell at a loss, then wholesale or "fleetail" the vehicle instead of retailing. Other over age new vehicles could be given to your BDC or Internet department to sell at cost. You should not retail at a loss and mini deals encourage this. Think about eliminating them from your pay plans. If you start paying 25% of the salesperson's total gross generated for the month, then the fact that some deals had a smaller gross will not be as noticeable in a report as they appear on an individual commission voucher. Mini deals throw your variable expense percent off. Creating a Report of F&I Gross Here's the biggest issue – you need to run a report from your F&I system of the payable gross profit for the month. You say that the F&I gross is wrong? This can't be! You must get accounting gross to match F&I gross then create the fields to correctly calculate payable gross. Your first step is to see if it's a problem by running a report that compares accounting gross to F&I gross. I've actually been at a dealership where these two amounts were exactly the same. I was shocked and when I asked the controller how he accomplished this – he merely responded, "why shouldn't they be the same?" Perfect answer! His system was that when the billers in the accounting office posted a deal they made sure that F&I had the same gross. If they didn't match, she stopped and questioned the sales manager or F&I manager and made sure sales and accounting were in sync on the deal's gross profit. Yes, we can say that "the sales department is never right," but that is scary. What if accounting missed a rebate and even missed submitting for a rebate? That profit is gone forever. See how commissions can affect profit? If this comparison of F&I gross to accounting gross is done weekly, then you can run it one more time after the month is closed and then run your commission report and pay commissions. That's another issue – I like doing sales commission on the 10th or 15th of the next month. During the month a salesperson gets a draw equal to their minimum wage or salary, we close the month with accrued commissions from the deal postings and settle out the commissions between the 10th and 15th. All your held deals should be cleared by then and any adjustments to gross should be done by this time. Trying to crunch in a sales commission payroll before you close the books is probably delaying your month end close past my 3 day guideline. If you have a lot of bonuses, then estimate those and adjust during commission payroll next month. Spilt Deals - This is easily done with a formula in Excel. The formula looks to see if there is a name in the salesperson #2 field and if so it splits the commission. Comm#1 column (D1) =IF(C2="",A2*0.25,A2*.125) Comm#2 column=IF(C2="","",A2*0.125) The formulas look to see if there is a salesperson's name in the Salesperson #2 field – if not, then it pays the full 25%. If there is, then it splits the amount (12.5%). Use the AutoFilter twice, once for the SLP#1 field, then again for the SLP #2 field – or use an Advanced AutoFilter using both fields as your criteria.

Salesperson Productivity report

Although the sales managers should be looking at this daily, each month, after all the deals have been booked, create a salesperson productivity report that shows, gross profit, gross profit average, number of deals, and F&I income. If you can, sort this by total gross profit.

Sales tax report

There are various ways to make your sales tax easier to report. The first step is to look at your reporting form and find what needs to be reported by type. For example if your state taxes vehicles by county, you can run a report from your F&I system of all the deals and amounts collected by county. If you must report leases separately, then consider setting up sales accounts for leases. For deductions, make sure your billers set up non-taxable sales in either a non-taxable sales account - or post them in their own journal.

Sales Tax Exemptions -Verify

After the sales tax report is completed, take the list of exceptions or deductions and verify paperwork for each deal.

Sales tax – pay

Most taxing agencies have created online methods for you to pay and report your sales tax. Find out if your state or providence has this available. Try to avoid paying what you collected; rather do a careful analysis of your sales minus deductions to come up with the actual amount due and then reconcile to the sales tax payable account.

Service Effective Labor Rate

If you sold every labor hour at your posted rate –then your effective labor rate would be the same amount each month – your posted labor rate. But we discount labor in our industry. For example a lube, oil, and filter might be .5 of an hour - or for a post labor rate of $100 an hour, $50 – but we sell it at $10.00 so that the total is only $29.95. That means only $20 an hour instead of $100! The reasoning is to get the customer to come in for the $29.95, but sell them much more business at the $100. If we sell 1 hour at the $20 an hour and 1 hour at the $100.00 – then add up the two amounts ($120) and divide by 2 hours. Your effective labor rate has been brought down to $60 an hour. This report takes all the labor for the month and divides by all the hours.

Service Open Repair Order Report - Analyze

This might be confused with Work in Process and they really are different - but the same! Work in Process means that the vehicle has work done already – often a portion of the flag time paid or parts issued to the vehicle. The Open Repair Order Report has all vehicles – even those who have not been touched yet. When you look at the Open Repair Order Report, you're concerned about why some vehicles have not been worked on yet. Be sure and filter out any that are still in Appointment or Estimate status and not officially open. You don't need to be concerned with those unless it is past the appointment date. If you have too many vehicles for your technicians to work on, then you'll want to focus on the areas that increase productivity like parts fill rate, training, or diagnostic tools. If productivity is already high, then you'll need to hire more techs and either increase your facility or add shifts.

Standard Entries

We already discussed standard entries in fixed assets, and vacation accrual, but in order to get to a 3-day close, you might need to accrue more items than normal. Be sure and use the advertising budget form to accrue from that as an AP entry instead of doing a standard entry.

State of Store Report – 20 Day Report

A few years ago I devised a 20-day report in Excel. For a few years and for a few special clients, I would dial into their DMS system and pull the data for the report and then email the Excel spreadsheet to them (for a fee-of course!) The purpose of the 20 day report was that it was done as close to the 20th of the month. Why did I pick that date? It's a perfect time because the prior month is closed (better be!) In addition, the current month should be all caught up, payables are paid, tax reports done, and we're at a rather stable point for looking at balances. In other words – there should be no excuses like there are for the financial statement (used vehicles or contracts in transit are high because of the "push" of deals at the end of the month.) I find the 20th – 22nd (depending on weekends) the perfect time to do your "State of the Store" - SOS report. It also gives you 10 days to save the store – instead of waiting for a poor financial statement and then trying to make changes. The first thing that you need to do is to design an Excel spreadsheet or use the one at ***SandraJerome.com – go to Contact – Free Stuff and click to download Super Controller zip file of spreadsheets.***

Titling – Regulations

With our increasingly mobile population, it is almost impossible to know all regulations for all states. What tools have you provided the sales and F&I departments to make sure they are informed about titling regulations? Many states now have online guides – or there are service providers that provide the information you need. Find out where in the process this responsibility lies and make sure they have what they need to log into the sites and use the services you are paying for. Make sure you have a written policy for the following; when can a title be process for a purchase, when are plates issues, registration provided, what trade-in documentation is required, when is sales tax required to be collected.

Unclaimed Property – Check

Most states have a website that lists unclaimed property. Check to see if there is any unclaimed property for any of your corporate names.

Unclaimed Property - Submit

According to a group of controllers that I recently spoke to, unclaimed property audits are on the rise as the states have found a new source of income. This audit can be expensive. The auditors will normally look at your journal vouchers and bank reconciliations – trying to find checks and balances that you wrote off to income instead of sending that money to the state. Do a self-audit yourself and make sure you're following your state's guidelines for submitting unclaimed property. You also need to find out the "why" of small credit balances. For example if credits are caused by over estimating license fees, then train the F&I managers to accurately calculate the fees. If they are adding some "padding" to the fees, get them to stop that. You can no longer keep it so why collect? Any buffer amounts should be part of your DOC fees.

Website - reconcile inventory

A typical dealership has a bunch of inventory databases; the DMS system, CRM, 3-4 websites, factory website, floor plan lender, etc. The biggest complaint I get from salespeople about the dealership's web site is that customers come in looking for a vehicle that is on the web site today – and it's been gone for weeks. How can you get your inventory "in sync?" First you need to decide that your DMS system is the "controlling" system for your vehicle inventory. I realize that you might think that it is "all messed up" and not worth the effort – but if you take my advice and use this as your ultimate basis for what you really own – then you'll have cleaner deals, closer grosses, and accurate costs for the sales department. If you buy a vehicle or sell a vehicle it will eventually reflect in this inventory record. It might have some timing issues, but by doing this exercise you'll find out what causes it and correct the process. Step 1. Download to Excel what you have in stock in your DMS system. It needs to include the following fields: VIN, stock#, date sold, accounting cost, year, make, model and color. Select only vehicles that have an accounting cost. Step 2. Take a physical inventory and enter into Excel. Most of the DMS systems have a product called vehicle bar coding that makes this task easier and you might want to invest in it. Step 3. Download from your manufacturer's website what they show you have in stock. Step 4. Download your outstanding floor plan vehicles. Step 5.

Download the vehicles from your web site. These download can be done in a few different ways. Do a search with a wildcard and get all your vehicles to display. Highlight the table, and from your Internet browser, select edit, copy. Then open Excel and edit, paste special, and select text. Do not select HTML or you'll have a lot of trouble removing the formatting. If it all comes in one column, click on that column and select Data, Text to columns. Next, the best way to do this HUGE purification is to put these downloads from each source into one spreadsheet with one tab for each source. Name each range, and then use a vLookup function to look up your DMS list (your master) on each of the different sources. Now it's time to find out why they don't all match. The most common reason why you have vehicles in stock in your DMS system that are sold might have to do with the process in your F&I department that closes deals and moves them out of inventory. That's why you pulled the field "date sold" in step 1 – if there is a date in that field, then it's normally sold in F&I, but if the cost still has an amount, the accounting office hasn't worked the deal yet. You might use this "date sold" field for further updates to your web site. If you have wholesaled or dealer-traded a vehicle and it's still on the list – that's not unusual. The office works these deals last due to the pressure to get retail deals on the DOC and pay commissions. Someone needs to update a field on your DMS system like "date sold" when the transaction happens. Create a better process for removing sold, wholesaled or dealer-traded units. Even worse are the units on your DMS list that are not on the web site. Find a better process for getting these fresh units up there quicker. Every day that goes by, you're missing an opportunity to sell this vehicle. What about vehicles that are on your website – but not on the DMS system yet? This process needs to be "in sync" – when they are added to the website, make an entry to bring them into the DMS system. Your sales managers and salespeople depend on this being accurate in the searches they do when looking for vehicles for their customers. If you're using CRM software, then that normally pulls from the DMS system – so again, you're not marketing the vehicles if they are not in your DMS system. Let's hope that someday, you'll have one system that connects all these sources, but for now – commit to keeping them "in sync" and you'll do a better job of merchandising the vehicles you own.

Website - update contact information

Once, a month, make sure dealership contact information is correct and all links function on your websites. Ask for a list of all dealership websites from your webmaster and personally go through each page and look at all the phone numbers, staff pictures, addresses, etc. Next click on the various links to make sure nothing is broken.

Wholesalers - Funds and titles due - monthly report to management

Putting your wholesales on the heat sheet is not an effective way to collect these amounts since neither salespeople nor F&I managers are involved. I prefer wholesales be posted to an open item AR account with the stock# as the reference.

That way monthly statements can be mailed to your wholesalers. To make the collection process a little easier; 1. There must be a way to notify the wholesalers that the titles are available and ready for pickup. The easiest way is to create a fax system to each wholesaler listing all the titles that are ready to be paid for. 2. There needs to be a set time when they come to get their titles. Pick times when your DMV clerk is there, and is not the time of the local auctions; for example Tuesday and Thursday AM from 10 – 11AM. The DMV clerk puts the amount due/ wholesaler's name and control# on the title and then puts in an envelope for each wholesaler. This can be left with the cashier or receptionist if you want. 3. Any wholesale title that passes the pick up day without getting paid for is put on the wholesale account heat sheet and a manager does the follow up when they don't pay. The best person is manager who wholesales your vehicles. This person should be given the faxes after they are sent to the wholesalers.
4. Make sure you are using a field in the inventory title to indicate if you have the title for a vehicle or not. Maybe your managers will delay wholesaling a vehicle until you have the title.

Wholesale Deals Sold At a Loss Report

Each month, prepare a report that lists all wholesale deals sold at a loss. Be sure the report shows the source of the vehicle (auction, new trade, used trade, and the person who appraised the deal.

Work in Process Labor - Reconcile

Work in Process is a closed loop. It has to balance each month. What does "closed loop"mean? Let's take the parts inventory for example. It is not a closed loop. There are actually three inventories; the GL amount, the "pad" amount, and the physical amount. Parts can be given to customers and technicians without being recorded on a parts ticket or repair order. On the other hand no technician can get paid without the office controlling it. It is a closed loop. It you pay them what is flagged on repair orders and the amount that is flagged is costed correctly, then it must balance to zero after adding up work in process. To reconcile the account, take the balance in the GL and compare to the cost amount of labor on the Work in Process or Open Repair order list. If you pay your technicians on booked labor through the end of the month – they should equal each other and no adjustment is necessary. Believe or not, I have many dealerships I know that these two equal each other. If they don't, there are some ways that you end up with large credits or debits in the account. Step 1 - Timing
To reconcile work in process you must first get your timing in sync. Do you pay your technicians weekly? If so, then work in process needs to be balanced weekly. Run your open repair list/work in process immediately after the last repair orders are updated for the pay period. As an added step at the end of the month, you'll need to accrue any technician earnings that have not been paid. For example if the month

ends on a Wednesday, and you paid the technicians through the last Friday, then you have three days of technician pay still due. Run your technician hours report and accrue that amount. If you are accruing 3/5th of a normal payroll amount, then you won't balance, but you can get close. I feel this whole entry is not required on a monthly basis – except at the end of your tax year and everything needs to be exact. I'd rather let the credit sit there and reconcile each week. If you pay your technicians through the end of the month, then you have an easy job to reconcile and no timing issue. Step 2 – Booked or Closed Time If everyone paid technicians on closed time, then you wouldn't even need a work in process account. There would never been "work in process" since you'd never pay a technician on unfinished work. The only work in process balance would come from the timing issue above and it would normally be a credit. Unfortunately, most dealerships pay technicians on booked time. This means when they finish a job or line on a repair order, they get flagged – or paid for that job. A repair order might have ten lines and a technician was paid on five of these lines, but it is still open. That is the reason why you need to run a work in process report. Step 3 – Your payroll distribution
I find most errors happen in the payroll distribution process. If your technician flagged 103.99 hours and they are paid $32 an hour, then the debit to the distribution code for work in process should be $3327.68. You can see that amount on most technician reports. If any other amount was paid to the technician and charged to work in process, then you'll be off. This would include bonuses, overtime, holiday, vacation or sick pay, shop time, or unapplied time. You need distribution codes for these amounts. Do not use a code that is the same as work in process. I find these errors by comparing the payroll register to the technician summary. Step 4 – Technician costing The amount of the cost on your technician summary must match the amount charged to work in process in step 3. If it doesn't, then the technician cost setup is wrong in the service department. There is no need to estimate the cost or add a "padding" to the cost. It should be exactly what you pay the technician. This is the 2nd most common error that I find. It is also found in step 3 while comparing the two reports. Step 5 – Hourly technicians Some dealerships let hourly technicians (apprentices) flag. These credits go into work in process, but the debit might hit other service wage expense. The opposite might happen; the full amount of hours the hourly technicians work might go to work in process – not just the amount they flagged. It is important to use two distribution codes for hourly technicians; one that hits work in process for their flag time and another that expenses their unflagged or guaranteed hours. Step 6 – Work in Process account audit Detail your work in process account. Is anything other than credits from repair orders or debits from payroll hitting this account? There should be nothing else except the accrual for weekly payroll if this applies. Step 7 – Back flags, partial flags Try to avoid both of these items. If you have a comeback, have the technician who made the mistake fix it right this time with no flag. There should be no partial flags. This is the incentive for a technician to finish a large labor operation and get the line completed. Step 8 – Teams. Teams should not affect work in process. All you're

doing is grouping the hours and then splitting the hours. If there are team bonuses and incentives that are not in the flag hours costs, then make sure step 3 is working properly and those bonuses are not a cost of sale, but wage expense. Instead of auditing technician by technician in step 3, group them by teams and then find them in the payroll journal and make sure the split amount equals the team amount. These might seem like a lot ways that work in process can be off, but once you have "closed the loop" your account will always balance with the work in process list and you can concentrate on reducing work in process and making sure vehicles get completed each day.

Tasks for the Close

This is probably the smallest close list that you've seen because if the daily, weekly, and monthly tasks are completed before the end of the month, then all you're really doing is preparing a financial statement – simple!

Car Deal Cutoff

Decide when you are cutting off the deals each month and email that information to all employees. For example if the 1st is on a Monday, then cutoff is noon on Monday. First thing in the morning, run a report of all the deals through the end of the prior month and find out which ones are in the office and what you still need. Next, send the missing list out to the sales and F&I manager and indicate that you need these in the office by noon to "post." Post is the most important word – you don't have to work these last few deals, you only need to post. If you've been doing a daily missing deals report, then there shouldn't be too many for them to turn in and for you to post. Remember that deal approval has changed dramatically in the past few years. Most approvals come instantly – so no longer do managers have to spend those first few days calling in deals. My daughter just bought a car and she was approved before she even took a test drive! Same with verifying insurance and payoffs – proof can now be obtained right while the customer is in the F&I office. No need to make those long phone calls when accounts can be accessed online.

Employee Add/Delete Report and Financial Stats

At the end of each month, the payroll clerk should prepare a month end report of all employees added and those terminated. After completing this report, prepare the Financial Stats worksheet showing total employees by department.

Final Entries – Accrue

This would be like a 2nd standard entry, but for those last minute items like bank charges, floor plan interest or special dealer cash that requires the final count of vehicles sold to accrue.

Reconcile DOC to Manager's Profit Estimate

Before you close the books get an estimate from each manager of the gross or net profit they felt was generated for the month. Compare the DOC to this amount and if it is off by a lot, give the DOC to the manager and ask what might be missing – for example a parts discount wasn't booked or floor plan credits weren't entered.

Financial Statement – Draft for Approval

Most dealers or GMs need to see the "numbers" before you accrue bonuses. This is considered a draft copy. To save yourself time, prepare this draft copy (before memo accounts is normally okay.) If you have a good idea of what the bonuses will cost you, put a note on the draft copy of the draft profit - less bonuses and a way for the person to approve this draft so you can complete the rest of the process without further approval (or delay.) Profit $63,343, Manager Bonuses est - $13,343 , Est net $50,000. Okay to submit X___________

Manager's Bonus Payroll – Calculate Accrue

Keep the bonuses in one spreadsheet with a tab for each manager. This makes it easy to email each tab to the manager. Save the spreadsheet by month and year to easily find them later. The DOC should have every line item that you need to accrue the bonuses and have an final income statement that reflects the data in the Operating Report. Compare the DOC to last month's - or run a comparative to check that all expenses are accrued and everything looks reasonable. If you currently do not have a functional DOC take the time to create a useful DOC as this will really help you with accruing bonuses and doing comparative analysis. Each employee paid a bonus should understand clearly how their bonus is calculated. If they don't know how they are being paid, how can they improve profits? Consider getting my Pay Plan Wizard to help you develop a pay plan that gets results.

Financial Statement Memo accounts

To make your close go quicker, create a memo worksheet by clerk. For example take the memo part of the personnel count and give that to your payroll clerk. At the top of the worksheet, provide instructions on how to adjust counts. Do this for your new/used inventory, etc. Ask the clerks to post/adjust their sections and turn this back into you. The best way to do memos - is to change your process so the memo accounts are right. Find out what causes them to get "off" and stop that from happening.

Financial Statement - Final Print

Some factories require that each franchise be kept separate and separate financial statement produced for each franchise. Others allow you to combine the balance sheets, but require a separate sales line for each franchise type. Make sure that when you do complete a final print of the financial statement that you save this as "final" in the filename and indicate the month and date. For example NissanFinalOct2013.xls.

Financial Statement - 20 group reporting form

Use the same method as memo accounts to speed up your 20 group reporting. If your 20 group needs a certain account broken down, find the person who can best do this and assign the task to them.

Financial Statement - Mail or email copies

Make sure everyone that you send the financial statement to each month really needs it. Banks normally only need quarterly and your CPA firm - only annually. Now that most manufacturers are going to Excel for the statement, ask if you can email the file instead of printing and mailing it.

Financial Statement – Transmit

Do you have a quick step guide to transmitting your financial statement? If so, add a "How to" to this task to make it easier to follow each month. You might want to take your DMS provider's most recent class on transmitting the financial statement. Often there is an easier way than the process you're using. They might have a quick guide that you can use for your "How to."

Yearly Tasks

Budget Expenses

In the fall of every year it is a good time to create a budget for next year's expenses. Expenses are difficult to budget. We have a bad month and sell fewer vehicles causing floor plan and advertising per unit to go up. Gross profit is low, and we paid more minis causing the sales commission variable expense percentage to increase. To complicate matters, towards the end of the month, the sales managers went into a panic and overspent the advertising budget in the fear of having a bad month. It ended up being a self-fulfilling prophecy – the increase in expenses coupled with the lower gross profit caused a double whammy and ouch – a bad month! How do you get managers to stick to their expense budgets when a slow month happens? First of all you need to create a budget that is not based on units – but reality. Let's move down the financial statement and create a budget. Sales commissions are easy. They are budgeted as a percent of gross profit after pack. It is the 2nd Most important item to perform an actual vs. budget on, after sales commissions. If your sales managers are having difficulty sticking to their budget, move the purchase order function to the controller and do not allow advertising without a purchase order. Advertising is not a variable expense. At the end of the month it does not go down like commissions if you sold fewer units. Consider it a fixed expense and budget a fixed amount. Floor plan should be an easy item to budget. You should have a days supply that you are going to carry and a cost for that days supply less any credits. Investigate the difference. Where does the overage come from? Units that were sold and not paid off in two to three days can cause excess floor plan interest. This might be a good month to create a report of date sold vs. date paid off for your new and used floor plan. The purpose of a budget is to find ways to reduce your expense by analyzing those processes that cost you money. When you do this report and find the delay, you might install a simple process like having the F&I managers submit the contracts to save money. I have a lot of dealerships that tell me that F&I managers do not have the time – but the average F&I manager does 100 deals. Most dealership experience 60-70% of those deals on the weekend leaving about a deal and a half a day left over during the week. Since most deals take an hour – that means five or six hours free to pull a contract – which only takes a few minutes. I realize that getting paid a day or two earlier for each contract only means a savings of $1500 - $2000 a month, but when you start plugging these holes in your expense budget, you'll start having better months. Delivery expense can be budgeted. Determine exactly what needs to be done to a vehicle to prepare it for delivery and make sure the expense matches the units you delivered. What else is in that account? Are there any credits? If you also charge this account with the gas for a demo ride, then use your showroom control software to tell you how many demos you had that month and calculate how much each demo ride's gas should be. Why go to this extent to budget a small account like delivery expense?

Because it can be one of those catch all accounts that hide waste and theft – especially if it is offset with credits. Pick this account this month to create a metric of how much per delivery and how much per demo and investigate the differences. Keep going down the expenses creating a budget for each. Don't use the easy method like your 20 group numbers, a percent of gross or last year's average. Every expense needs to have a budget based on real amounts. If you have three lot people in other salaries, then the budget should be exactly their wages. Any variance points to a problem like overtime that needs to be investigated. Yes, this is a hard chore but start with a few accounts and continue from there. Have a detailed policy of what goes into that account and how you budget the amount and who needs to be alerted when you have a variance.

Budget and Forecast – Gross Profit and Sales

Create the budget and forecast for next year for gross profit and sales. The new and used departments forecast units and budget gross profit average. If you are in a seasonal area then you'll need to forecast units for each month. For example, you sell a lot more used in Minnesota in August than you do in April, but April is a good month in California. I find that forecasting units is one of the hardest things to do in a budget. You have a few resources; previous year's performance, 20 group averages, economic forecasts. I did a search for a make and forecasts and found that this import plans on capturing another ½ of a percent of the market the next year. That makes your job easy; take last year and increase by at least ½ of a percent. Not very much - but a starting point. If you forecast any more than that – you need to have a plan. Let's say you're going to beef up your Internet department or increase the activity of your BDC from just making service department appointments and following up on non-buyers – to campaigning for new sales. If that plan is going to add 5 more new a month, then forecast for it, but make sure you note why it is going up. Let's do our budget for gross profit. For new and used, this starts with an order mix report. You need to go back over the last two years and pull all your new vehicle sales. Find your gross profit per unit average and number sold. Now you need to determine what will be your inventory mix. Indicate by model – how many will you carry in stock each month. From that, you can determine how many you will sell and the corresponding gross profit average. This is an important step because in our expense exercise we'll need to know what you will carry each month to determine floor plan expense. I find that few dealers have an order mix report created and they certainly do not budget for it. This should be a chore for your inventory manager – and if you don't have an inventory manager, a project for you controller, sales manager or Internet/BDC manager. Forecasting service hours is a little easier. Take the number of technicians and multiply times the number of hours each month and their productivity rate. To do the gross profit, take your effective labor rate and subtract your technician average cost per hour. I find that parts is normally a percentage of service gross profit since most of their income comes from repair orders. This varies by manufacturer and in my example, the rate is 75% of labor gross profit. How does budgeting and forecasting create profit? The key is in performing an actual comparison. When your month is over, do a detail comparison of what your actual results were to budget and forecast. What happened to miss your budget? Let's say that we didn't sell the 40 new that we had planned. Did we install that campaign in the BDC? What if we missed the gross profit average? Did we carry the right order/mix? What if we missed the labor hours? Did we turn away customers? If we got the hours but missed on service gross – did we hold the effective labor rate? Each of these items that were budgeted and forecasted can point to various profit points that need investigating.

Bad Debt Estimate

Adjust your bad debt estimate based on last years' write offs and your current accounts receivable balance.

Contracts – Double Check Dates

Once a year, check the ending contract date for all your contracted vendors. Many DMS contracts "auto-renew" if you don't terminate it within the guidelines and you could get stuck for another 5 or 7 year term.

Corporate Annual Report and Minutes

Update the corporate book with any minutes, have your annual meeting and file your annual report and pay any fees.

Demo Plate report

Most floorplan reports indicate which vehicles are demos and can easily be used for the basis of your demo report - but if you have used demos and they are not floored, then you might need to manually add them to the demo report. A demo report should indicate who is driving the vehicle, how many miles, and demo plate attached. Make sure you list the other demo plates on the report and who is in control of the plate. That way you'll order exactly how many plates you need each year and not pay for plates you don't need (or don't have anymore.)

Employee Job Description Update

You should have folder on your dealership service that holds all the job descriptions. One of the first questions I ask when a manager or dealer is disappointed in the performance of one of their employees is, "what does it say in their job description?" I knew a dealer that was furious that the controller had not reconciled the bank statement in over a year – but sure enough that was not in his job description. Many of you might argue that everyone knows that making sure the bank gets reconciled is a basic function of being a controller – but for this poor guy, the bank was reconciled by his CPA firm in his prior job. I've had controllers that didn't know they were supposed to follow up on rebates, reconcile floor plan statements, or even make payroll deposits. A key to a job description is to make sure these holes don't happen and most importantly, that you get what you want from that person. If you hired me as your controller, your objectives might be for me to optimize your computer system and work with the other managers on their department's profitability. I'd be shocked to find out that you expected me to file, cashier and answer the phone – but I know many controllers that have that as part of their daily functions. Before you hire another manager, grab a copy of their job description and look at what is expected of them. Don't have a job description for each position? **Our Super Controller II – Expenses and Employees** has lots of them included in Word format. My job description format has four sections; The position/department, general description, duties and responsibilities, and skills/physical demands. Let's look at how each of these sections helps you prevent disappointments and make more money. The position section is more important than you think. It clearly defines the department that the person works for and who they report to. I can't tell you how many times in my career that I didn't really know who I worked for. When I was an assistant to the F&I department, I had six or more bosses; the F&I managers, the sales managers, the general manager, the F&I director and of course the controller got stuck with me since I was a female. I know I didn't make effective use of my time since I was often used to fetch a sandwich for the sales managers. I like to think of the general description as a mission statement for the employee and is best when it is three sentences or less. It is the Duties and Responsible section that should make you money with this employee. What are your objectives for this position? For example in my job description for an Internet manager, item six is, "Review billings from web site providers and reconcile any leads or development costs to actual logs." This expense cutting function normally saves the dealership thousands of dollars. Don't expect that your Internet manager is doing this automatically. The last section that defines skills and physical demands is the one that protects you legally. How can you fire someone for a bad driving record if it was never defined that their job required them to take the deposit to the bank or perform dealer trades? If you need this person to drive a company vehicle, then make sure you include that requirement. If they need to be able to lift file boxes, then make sure they can pick up 50 pounds if that is part of their job. You might want to run that section by your legal advisor just to make sure you have the correct requirements or wording for the position. No matter what format you use for a job description make sure you create one for every employee – and make sure the people involved get a copy. If you hire a service advisor – give them a copy and give a copy to the service manager. I'd prefer the service manager edit the description before you even run the advertisement for the position. As you can imagine, I have added to the F&I manager job description item12; "Submit the contracts to each lender with all the proper documentation to ensure prompt payment and maintain a record of when and how contracts were submitted."

Employees- Calculate Turnover Rate for the Year

Based on an industry guide of 25%, if your dealership is experiencing a higher turnover rate, then the cost adds up. In the example below, this dealership had 173 W2 forms and an average of 130 employees. That is a 33% turnover ratio. That means they had 10.5 employees higher than other dealerships. Turnover costs are estimated by experts at 75% of annual salary. This is the cost of training, loss of productivity, and simple things like adding the employee to the payroll system and printing new name tags. If you divide your total W2 wages by the number of employees and then take 75% of that times the 10.5 extra turnover that you have – it can add up to a huge number.

Filing – Year End

Every accounting office runs out of room. You can cut down the amount of documents that you store by printing less and saving more to your server (see scanning tips.) The first step is to get a record retention guide from your CPA firm or state association. I think those fancy bankers boxes are a waste of money – especially when you're going to destroy them anyway in a few years. As the parts department to save boxes of average size for you. When it is time to rotate out old files, put the stuff in an old box and label "Destroy 2015" and "Car Deals A-M, 2007." Since so much information is now scanned and on the computer, you shouldn't need to be going in and out of these boxes too much.

Holiday Schedule

Determine holiday schedule for next year and update company calendar.

Income Tax – Return or Extension

File tax return or extension before due date.

Labor Rate – Apply for Increase

Most manufacturers allow you to apply for an increase your labor rate periodically. See if you are eligible to make an application and apply. Consider increasing your Customer Pay each year based on increases in technician payroll and benefit cost increases.

Lemon Law – Guidelines and Fees

Update your guidelines for the Lemon Law and pay any required fees.

LIFO Estimate

Make any LIFO estimates (contact your CPA) before completing your last financial statement for the year. This estimate normally have to be on that statement.

LIFO Prepare

Before the year ends, find out what will be required for LIFO and be ready to write those reports or create copies on the last day of the year – or a few days into the year (after you final inventory is processed.) Save any price tapes that you might need for parts inventory.

Other Taxes – Fuel, Tire, Luxury

Determine filing requires and fees for other taxes like fuel, tire, luxury, etc.

Payroll – order end of year forms, W2

Normally in the fall, the W2 forms are available for order. Check with your payroll provider or DMS company for the correct format and then run a report on YTD earnings to determine how many you need. This is a good time to check the active/termination status. If someone has no YTD earnings, then they were probably terminated and the proper fields not checked. Get that updated now before you end up printing a blank W2. Find out any state forms that are required to report earnings and it is a good time to order 1099 forms at the same time.

Payroll – Update Tax Tables

Obtain next year's tax tables and update rates for federal, state and local taxes along with exemptions. Update unemployment, FICA and Medicare rates.

Payroll – Update Demo, EIC and Sick Pay

Update the amount that you need to take out for demos, and any 3rd party sick pay and EIC amounts.

Parts Inventory, reconcile Physical to Accounting

It is rare that the value of parts counted during physical inventory is equal to the totals in GL balances. It is impossible for parts inventory to be higher than accounting – why? Parts don't multiply. If you have more parts in stock than shows in accounting then you have an error in your accounting process. Somehow parts are coming in stock without an accounting entry (take off tires, etc) or appreciation isn't being taken. A shortage in the parts department can come from incorrect accounting or theft. Here is a checklist to find shortages: Check the parts reconciliation form. Are the items subtracted or added correctly (it's common to enter them in backwards.) Look at the gross profit percentages on the financial statement – are they higher than other dealers of the same make? When a dealer purchases a part from the outside that is for their make of vehicle, they must make a cost adjustment on the invoice to debit parts inventory for the tape amount and debit cost of sales for the markup from the other dealer. Is this being done? Are there deletions being made on the parts pad? This normally indicates theft of the part. Look at the override report – are they zeroing the cost on tickets? This normally indicates theft of the part. Has appreciation been taken? See the section on appreciation. In the accounting records investigate all checks and journal vouchers to parts inventory and any items from non-traditional sources (a debit to parts inventory and credit to cash sales.) This normally indicates theft in the accounting office.

Personnel Budget

Each year determine your staffing budget for the accounting office and ask other managers to do the same using the spreadsheet in the free downloads. Detail personnel by name, normal wages and forecasted overtime. In the spreadsheet, the yellow portions are what you forecast – and the green cells are based on prior history. At the bottom is an "over/under" to use to tweak.

State Dealer License Renew

Renew your state dealer license and any other licenses like insurance, broker, or salesperson licenses. Pay any fees for licenses or safety inspection (fire) fees.

Vacation Accrual

Vacation Accrual is an annual estimate that should be a part of your standard entries each month. You should not directly expense vacation time each month – that makes your expenses jump up high in the popular summer months and distorts the actual expenses for that month. If you do a vacation estimate once a year, then the task can be easier especially when you're comparing to the previous year's results. I found a free spreadsheet on the Microsoft website that does a great job of tracking vacation, personnel and sick leave. It is in the free downloads. I had do some modifications to make it ready for each year, but now it's all ready for you to use. The best features is that it creates a year to date summary page and I've added a few columns to create the amount you need to accrue for each employee.

Vacation Schedule

As a controller, I hated the summer. Not only did sales increase, but my office staff and other key personnel wanted to take vacations. This meant special vacations checks and trying to figure out how much pay they had coming. Add this annoyance to the lack of coverage and the inconvenience of somebody being gone that might have the answer or documentation that I needed to complete a certain chore. What are the steps that you can take to make this stressful vacation time go better this summer? 1. Schedule Vacations. Now might be the time to get a dealership-wide calendar. Nothing is worse than have someone on vacation the week you've paid for an expensive consultant or trainer to be there! There are a few options; share your calendar in Outlook (if you're on Microsoft Exchange,) get a Google Calendar buy a DMS or CRM that has a calendar or scheduler (have you looked at DealerTeam?) Even the the old-fashioned dry erase board works, (although risky, I once leaned against one and smeared April!) 2. Cross Training. Vacations are a perfect time to do a little cross training. As soon as someone had told me they wanted to take a vacation, I'd remind them that their "how to" book needed to be up to date and that they would need to train someone to do their job while they were gone. I'd get their input on who would be the best person and then call that person in while the potential vacationer was still in my office. Of course the typical "what's in it for me" person would complain that they didn't have time to cover someone else's vacation – but it was better that they do that in front of the person whose vacation they were ruining instead of me! If that did happen, I'd express my disappointment and suggest to the vacationer that they pick another time when it would be better for our fussy person – and that normally did the trick and got things settled quickly. I made sure that everyone working for me understood that advancements and raises depended on how much they knew how to do, so it encouraged my staff to learn as much as possible during these vacation opportunities. What about those difficult other department positions like F&I, Service Advisor, Parts Counterperson? As a controller, I would volunteer to help out whenever I could. This is a perfect opportunity for controllers (or even GM and dealers) to learn other positions. Your first step when you get notified of an operational vacation coming up is to take the DMS training. For example if you've always wanted to learn more about the service department, take the advisor training. Next, you need to "shadow" the person you'll be covering for to see what their job involves. I've had people ask me over the years what was one factor that helped me do my job better. Many thought it was my accounting degree, or CPA certificate – but in reality, it was the fact that I had done the job of so many other positions and understood what it involved. 3. Track and Document. In addition to keeping your Vacation Track updated, you also need an official vacation request leave form completed that has the employee indicate how much leave they "think" they have coming so there is no last minute confusion on how much of a vacation check they are owed. Check that form today to make sure that item is on the form.

W2 Forms - Print

Print year end W2 forms. Provide to the employees before the due date and then mail or transmit the copies to the states and SS office.

Year end 1099 Statements

Print and mail year end 1099 statements to both the payee and to the state/federal agencies as required.

Year-end Payroll Tax Reports

Complete all year-end tax reports as required.

Year End Used Vehicle Write-Down

Find out from your CPA the guidelines for the year end used vehicle write-down and make sure you save a copy of the guide – or online estimate that you used to create the write-down.

Year End Review

Get a checklist from your CPA of any items that you'll need for the year end review; expense accounts that need detail, fixed asset purchases, and account details. Find out any payroll that you need to process for the owners.

Not Fair and Goodbye

It might not seem fair that I am now selling this guide for a lot less than before. Some of you might have paid for or attended the Super Controller webinars or live seminars at a great cost; but I haven't sold these guide for $295 - $595 for the past 4-5 years while working on installing DealerStar DMS and you've had those years to be ahead of the rest. One thing that I have to admit about our industry is that it isn't fair. My husband spent a career as a technical writer for the nuclear industry and Department of Defense. His jobs were very fair; pay was based on grades, scales, level and merit promotions. He had lots of overtime, vacation, sick, and holiday pay. My job was the opposite. I once hacked into the systems of the other dealerships in our group and found that other controllers at dealerships making less profit with less qualifications and abilities than me made more than I did. Later, in another dealership I found that my fellow managers got year-end bonuses that were 5-6 times more than mine; even when their departments were losing money and they were doing a terrible job.

Overall, male CFOs made more than females CFOs and when I worked in F&I, my male "partner" made twice what I did with the same hours and less production than mine. If you want to watch a funny video, watch this inequality video. It is from a Ted talk and a study of some monkeys being given a cucumber for performing a task. But then they give one monkey a grape for a reward and the monkey that was perfectly happy earning a cucumber sees the other one get a grape and throws the cucumber back and slams her hand down, angry.

https://www.youtube.com/watch?v=meiU6TxysCg

Dealers often wonder why employees steal. They merely want their grape. I had a dealer that loved to have me come into his office and tell him another way that I found where employees had stolen cash, cars, etc. in my long career. I did a fantastic seminar at AICPA called "Catch me if you Can" with Dave Jarrett of Crowe (a leading automotive CPA firm.) The AICPA refused to record it because they worried that attendees would use those methods. I used to keep my dealers on their toes by saying, "I am an honest person, but if I did decide to steal from you; you'd never find it."

I wrote about this in my blog - https://sandijerome.blogspot.com/2026/01/catch-me-if-you-can.html

I was good with computers and smart. But fortunately, I was raised with a strong sense of right and wrong. I've never sued anyone and never been sued – never been fired and to the end stayed friends with most of my dealers. In fact, the last dealer that I worked for – when I'd come back to consult with him, I'd stay in his house.

I didn't get my grape most of the time but learned to be happy with the cucumber. When I changed my "wants" to "needs" in life; I was able to accept things better and find other ways to get "paid." One of those ways was to invest in myself. When I had free time at work, I spent it learning the computer system which helped me when I left to be a computer consultant and eventually design a DMS. I went back to school; got my degree and studied for the CPA exam. I dove into the other departments and learned the other managers' jobs. I learned more and more about his amazing industry. I had a friend that worked at a bank. I envied her fair pay, nice plush office and "banker hours" but she talked about sitting and watching the clock until quitting time; bored. I was never bored and the only reason for clock watching was that I was always worried I wouldn't get it all done on time. This industry has such a sense of urgency and is thrilling.

Enjoy the guide and your journey. My thanks to everyone who made my journey one of the most exciting and rewarding careers in the world. As I move into my next job of becoming a writer and author, I will think of you all often.

Sandi Jerome, 2026

https://www.SandraJerome.com/

Made in the USA
Coppell, TX
05 November 2026

11314458R00188